NOTE.

IT has been my aim, in collecting and arranging the
Dr. Gray's large correspondence, to show, as far as possible in his own words, his life and his occupation. The greater part of the immense mass of letters he wrote was necessarily purely scientific, uninteresting except to the person addressed; so that many of those published are merely fragments, and very few are given completely. I have made no attempt to estimate his scientific or critical labors, for they are sufficiently before the world in various printed works; but something of the personality of the man and his many interests may be learned from these familiar letters and from even the slight notes.

Dr. Gray began an Autobiography, but went no further than to give a brief sketch of his early life. This fragment is placed, with some notes illustrative of the early conditions in which his youth was passed, at the beginning of the work.

It is owing to the kind assistance of many friends that the Autobiography and Letters are thus presented; among whom should be especially mentioned Professors C. S. Sargent and Charles L. Jackson, Dr. W. G. Farlow, Mr. J. H. Redfield, and Mr. Horace E. Scudder.

J. L. GRAY.

BOTANIC GARDEN, CAMBRIDGE,
July 1, 1893.

CHAPTER I.

AUTOBIOGRAPHY.

1810-1843.

MY great-great-grandfather, John Gray, with his family, among which was Robert Gray, supposed to be one of his sons, emigrated from Londonderry, Ireland, to Worcester, Mass., being part of a Scotch-Irish colony.[1] The farm they took up was on the north side of what is now Lincoln Street.

Robert Gray, my great-grandfather, died in Worcester, January 16, 1766. He married Sarah Wiley[2] about the year 1729. They had ten children; the eighth was Moses Wiley Gray, my grandfather, born in Worcester, December 31, 1745. About the year 1769, he married Sally Miller, daughter of Samuel and Elisabeth (Hammond) Miller, of Worcester, and removed to Templeton, Mass. About 1787 he removed to Grafton, Vermont, where his wife died in 1793. In 1794 he removed to Oneida County, N. Y., and settled in the Sauquoit Valley,[3] where he died from injuries received from the fall of a tree, May 8, 1803.

My father, Moses Gray, was the youngest of the (eight?) children of his mother. There were three half-brothers and a half-sister by a second wife, born in Oneida County, none of whom survived my father. He was born in Templeton, Mass., February 26, 1786.[4] He was therefore in his eighteenth year when his father died. He used to say that he had only six weeks of schooling; whether before or after his father's death I am ignorant. But soon after that event he was apprenticed to a tanner and currier (Mr. Gier) at Sauquoit, in whose employment he must have been for a part of the time after he came of age, for I was born in a little house which had been a shoe-shop on the premises of the tan-yard.

The fact of being born supposes a maternal ancestry. July 30, 1809, my father married Roxana Howard. She was born in Longmeadow, Mass., March 15, 1789; was a daughter of Joseph Howard, who was born in Pomfret, Conn., March 8, 1766, and of Submit (Luce) Howard, born at Somers, Conn., April 3, 1767;[5] and he was the grandson of John Howard of Ipswich,[6] Mass., and of Elisabeth Smith, of the same town. He was the descendant of Thomas Howard, who, with his wife and children, came from Aylesford (or Maidstone), Kent, in the year 1634.

Letters of Asa Gray

(Volume I)

Asa Gray

Editor: Jane Loring Gray

Alpha Editions

This edition published in 2022

ISBN : 9789356718289

Design and Setting By
Alpha Editions
www.alphaedis.com
Email - info@alphaedis.com

Contents

My mother came with her parents to Oneida County and the Sauquoit Valley when only a few years old.[7] Her father there joined a company which set up an iron-forge. One of the early pieces of work of its trip-hammer was to forge off three of my maternal grandfather's fingers. This appears to have qualified him to be the cleric in charge, or manager, of the office and store of the Paris Furnace Company, which established a small iron-smelting furnace on the Sauquoit, two and a half miles above the village of Sauquoit, in a deep and narrow valley which had the name of Paris Furnace Hollow, now called Clayville, the furnace long since having disappeared, a natural consequence of the exhaustion of the charcoal furnished by the woods of the surrounding hills. My earliest recollections are of Paris Furnace Hollow, for not long after I was born, as aforesaid, in Sauquoit, on the eastern or Methodist side of the creek, on the 18th of November, 1810, my father and mother removed to Paris Furnace with me, their first-born, and set up a small tannery there. Of this I retain some vivid recollections, especially those connected with the first use to which I was put, the driving round the ring of the old horse which turned the bark-mill, and the supplying the said mill with its grist of bark,—a lonely and monotonous occupation.[8] I was sent to the district school near by when three years old; and I either remember some of my performances of that or the next year, or have been told them in such way as to leave the matter doubtful.[9] My earliest distinct recollections of school are of spelling-matches, in which at six or seven years I was a champion.[10] There was a year or two of early boyhood in which I was sent to a small "select" or private school, taught at Sauquoit, by the son of the pastor of the parish; a year or two following, in which I was in my maternal grandfather's family, near by, as a sort of office-boy; and at the age of twelve, or near it, I was sent off to the Clinton Grammar School, nine miles away, where I was drilled after a fashion in the rudiments of Latin and Greek for two years, excepting the three summer months, when I was taken home to assist in the corn and hayfield. For my father, buying up, little by little, lands which had been cleared for charcoal, had become a farmer in a small way, an occupation to which he was most inclined. So about these times he sold out the tannery and bought a small farm nearer to Sauquoit, mainly of the land which my maternal grandfather had settled on, including the house in which he had married my mother. To it he removed, and there resided until he bought out an adjacent small farm in addition, with an old house very pleasantly situated, which he rebuilt and lived in until after I had attained my majority. But soon after that he bought a small farm close to the Sauquoit village on the western or Presbyterian side, hard by the meeting-house the family had always attended. There my father indulged his special fancy by rebuilding another old house, and the place, after his death, and, much later, after that of my mother, fell to my eldest brother, who still possesses it.[11]

I am not sure, but I think it was after two years of the Clinton Grammar School that I was transferred to Fairfield Academy.[12] Fairfield, Herkimer County, lies high on the hills, between the West and East Canada creeks, seven miles north of Little Falls. I went there first in October, 1825, the date I fix by that of the completion of the Erie Canal. For that autumn, I think in November, I walked one afternoon, along with some other students, down to Little Falls to see there the arrival of the canal-boat which bore the canal-commissioners, with the governor, De Witt Clinton at their head, on their ceremonious voyage from Buffalo to New York city. It reached Little Falls near sunset, and we walked to Fairfield that evening. The reason for my being sent to Fairfield Academy was that the principal of the academy was Charles Avery, uncle of my companion from infancy, Eli Avery, of our town, who died two years ago, who had been educated by the help of Eli's father, Colonel Avery, one of the owners of Paris furnace. Charles Avery several years later took the professorship of chemistry, etc., at Hamilton College, lived to over ninety, I think, and through all his later years seemed to be very proud of having been my teacher. I cannot say that I owe much to him, even for teaching me mathematics, which was his forte. My capital memory allowed me to "get my lessons" easily, and that sufficed; and I had none of the sharp drilling and testing which I needed. He lingers in my memory in another way. He was sharp at turning a penny in various ways; among them, he for the first year and more jobbed the board of his nephew Eli and myself, who were chums, paying for it in cooking-stoves and the like from Paris furnace, in which through his brother he had an interest, and boarding us round, from one house to another (we had our room in the academy buildings) until the stove which cooked our dinner was paid for. Sometimes our fare was good enough; but one poor widow, who took us in her turn, fed us so much upon boiled salt cod, not always of the sweetest, that the sight of that dish still calls up ancient memories not altogether agreeable. I think it was not at that time, but at a somewhat later date, and with less excuse, that we mended our diet upon one occasion, one winter's night, by carrying off the principal's best fowls from the roost, skinning them, as the most expeditious and neatest way, and broiling them in our room as the pièce de résistance, for they were tough, in a little supper we got up.

I here recall a favor which Mr. Avery did me. A year or two after I had taken my M. D., my dear old friend Professor Hadley, of Fairfield Medical College, who had been filling the place at Hamilton College pro tem., made me a candidate for the professorship there of chemistry, with geology and natural science. But my old teacher, Mr. Avery, an alumnus of the college, entered the lists and carried the day. I wonder if I should have rusted out there if I had got the place.

I must go back to say something of my omnivorous reading, which was, after all, the larger part of my education. I was a reader almost from my cradle, and I read everything I could lay hands on. There was no great choice in my early boyhood. But there was a little subscription library at Sauquoit, the stockholders of which met four times a year, distributed the books by auction to the highest bidder (maximum, perhaps, ten or twelve cents) to have and to hold for three months; or if there was no competition each took what he chose. Rather slow circulation this; but in the three months the books were thoroughly read. History I rather took to, but especially voyages and travels were my delight. There were no plays, not even Shakespeare in the library, but a sprinkling of novels. My novel-reading, up to the time when I was sent to school at Clinton, was confined, I think, to Miss Porter's "Children of the Abbey" and "Thaddeus of Warsaw"—the latter a soul-stirring production, of which I can recall a good deal; of the former nothing distinctly. One Sunday afternoon, of the first winter I was at Clinton, I went into the public room of one of the two village inns, where half a dozen of the villagers were assembled; and one was reading aloud "Quentin Durward," which had just appeared in an American (Philadelphia) reprint. This was my introduction to the Waverley novels. The next summer, when at home for farm work, I found "Rob Roy" in the little library I have mentioned, took it out and read it with interest. In the autumn, when I went back to school, some college (Hamilton College) students were boarding at the house where I boarded and lodged. One of them, seeing my avidity for books, introduced me to the librarian of the Phœnix Society of the college, which had a library strong in novels, which I was allowed, one by one, to take home for reading. I suppose that I read them every one.[13]

It was intended that I should go to college, and my father could have put me through without serious inconvenience; but he was buying land about this time, and he persuaded me to give up that idea and to go at once at the study of medicine, which I did, in the autumn of 1826, beginning with the session of 1826-27 in the medical college (of the western district), then a flourishing country medical school at Fairfield. I had already attended its courses in chemistry, given by Professor James Hadley (father of Professor James Hadley of Yale College, then a lad), my earliest scientific adviser and most excellent friend. I had a passion for mineralogy in those days, as well as for chemistry. The spring and summer of 1827 I passed in the office of one of the village doctors of Sauquoit, Dr. Priest, and on the opening of the autumn session returned to the medical school at Fairfield.[14] That year, in the course of the winter, I picked up and read the article "Botany" in Brewster's "Edinburgh Encyclopædia," a poor thing, no doubt, but it interested me much. I bought Eaton's "Manual of Botany,"[15] pored over its pages, and waited for spring. Before the spring opened, the short college session being over, I became a medical student, after the country fashion, in the office of Dr. John F. Trowbridge of Bridgewater, Oneida County, nine miles south of my paternal home; continued there for three years, except during the college sessions, where I attended four annual courses before taking my degree of M. D. at the close of the session of 1829-30.[16] The fact will appear, which I did not reveal at the time, that I took this degree six or seven months (I passed my examination, indeed, eight or nine months) before I had attained the legal age of twenty-one. But I looked older, and was in fact such an old stager in the school that no one thought of asking if I was of age. That degree gives me my place high enough on the Harvard University list to entitle me to a free dinner at Commencement.

I have mentioned my interest in botany as beginning in the winter and out of all reach either of a greenhouse or of a potted plant. But in the spring, I think that of 1828, I sallied forth one April day into the bare woods, found an early specimen of a plant in flower, peeping through dead leaves, brought it home, and with Eaton's "Manual" without much difficulty I ran it down to its name, Claytonia Virginica. (It was really C. Caroliniana, but the two were not distinguished in that book.) I was well pleased, and went on, collecting and examining all the flowers I could lay hands on; and the rides over the country to visit patients along with my preceptor, Dr. Trowbridge, gave good opportunities. I began an herbarium of shockingly bad specimens. In autumn, going back to Fairfield for the annual course of medical lectures, I took specimens of those plants that puzzled me to Professor Hadley, who had learned some botany of Dr. Ives of New Haven, and had made a neat herbarium of the common New England and New York plants, which I studied carefully that winter. At Professor Hadley's suggestion I opened a correspondence with Dr. Lewis C. Beck of Albany,[17] who was the botanist of the region. The next summer I collected more easily and critically. The summer after, I think, or probably the summer of 1830, I had an opportunity to make a little run to New York, being sent by Dr. Trowbridge to buy some medical books, driving in a one-horse wagon, with my own horse, ninety miles to Albany, thence by steamer to New York over night; one night there, and back next day by boat to Albany, and so driving back to Bridgewater in company with a man of business who joined me in this little expedition. I stopped to see Lewis C. Beck at Albany Academy; there I first saw a grave-looking man who I was told was Professor Henry, who had just been making a wonderful electro-magnet. I had procured from Professor Hadley a letter of introduction to Dr. Torrey, whose "Flora of the Northern United States," vol. i., was our greatest help so far as it went, and which on that journey I bought a copy of. I took also a parcel of plants to be named. Finding my way to Dr. Torrey's house in Charlton Street with my parcel and letter, I had the disappointment of finding that he was away at Williamstown, Massachusetts, for the summer. It was not until the next winter that at Fairfield I received a letter from Dr. Torrey, naming my plants, and inviting the correspondence which continued thence to the end of his life.

In addition to Dr. Hadley's summer course of lectures on chemistry, Dr. Lewis C. Beck used to come and deliver a short course of lectures on botany. He gave this up the year in which I received my M. D., so Professor Hadley invited me to come and give the course instead. The course was given in five or six weeks, beginning in the latter part of May. I prepared myself during the winter, gave this my first course of lectures, cleared forty dollars by the operation, and devoted it to the making of a tour to the western part of the State of New York, as far as Niagara Falls, Buffalo, and Aurora,—a dozen or more miles off,—where I visited an uncle, my mother's brother, a well-to-do country merchant, also a chum, Dr. Folwell, in Seneca County, high up between the two lakes, where I passed a week or two; thence to Ithaca, and across the country by a stage-coach back to Bridgewater. I hardly know what I did the next autumn and winter, but in early spring a Mr. Edgerton, a pupil of Amos Eaton, at Troy, the professor of natural sciences at the flourishing school of Mr. Bartlett at Utica, died. I applied for the vacancy, received the appointment, and for two or part of three years, minus a long summer vacation, I taught chemistry, geology, mineralogy, and botany, to boys, making with the boys very pleasant botanical excursions through the country round. My first summer vacation, if I rightly remember, was in cholera year, the disease being very fatal in Utica. About the time it made its appearance in New York I started off from Bridgewater, taking a little country stage-coach down the Unadilla to Pennsylvania; visited Carbondale and made a collection of calamites and fossil ferns; thence by stage-coach through the Wind Gap to Easton; thence out to Bethlehem, where I passed a day with old Bishop Schweinitz,[18] gave him a Carex which he said was new, but I told him it was Carex livida, Wahl. (and I was right); back to Easton; thence up to Sussex County, N. J., collecting minerals (Franklinite, etc.); thence to adjacent Orange County, N. Y., collecting spinelles, etc., as well as botanizing; thence down to New York early in September; there I met Dr. Torrey for the first time, and we took a little expedition together down to Tom's River in the pine barrens, and back to New York in a wood-sloop.

The next year, in the spring, Dr. Torrey went to Europe, sent to purchase apparatus for the New York City University, then just established. He engaged me to go that summer to collect plants in the pine barrens of

New Jersey, he to take the half of my collection, paying what would be required to defray my very moderate expenses in the field. I found afterwards that these plants went to B. D. Greene and his brother Copley, then abroad and full of botany; and I have encountered them, *i. e.*, the specimens, in various places, especially in Herb. De Candolle, as "Coll. Greene." I got down, I hardly now know how, to Tuckerton on the Jersey coast, botanized at Little Egg Harbor, Wading River, Quaker Bridge, and Atsion. While at Quaker Bridge my loneliness was cheered by the appearance of a fine-looking man, who came in a chaise, looking after some particular insect. It proved to be Major Le Conte.[19]

The next winter at Bartlett's school. In the spring went north to Watertown; visited Dr. Crawe, botanized on Black River, made mineralogical excursions, and back to Utica via Sackett's Harbor (lake to Oswego, and canal to Utica). After the spring term of school there—I think it was that year, but am uncertain—I took through the summer Professor Hadley's place at Hamilton College, Clinton; gave for him a course of instruction in botany and mineralogy. This, I have reason to think, was a ruse of my good friend, who wished me to succeed to that professorship, which he was on the point of resigning. Fortunately, Charles Avery, my old academic preceptor, became a candidate and secured the election.

These years are a good deal mixed up, and I cannot settle their dates nor the order of events. Only I know that the next autumn I got a furlough from the school until toward the end of winter, that I might accept Dr. Torrey's invitation to be his assistant during his course of chemical lectures in the Medical School, and at his house in the herbarium, living with him, and receiving eighty dollars as pay. This I can fix as the winter of 1833-34 or 1834-35. The first century of my "North American Gramineæ and Cyperaceæ" was got out that winter, and it bears the date of 1834.[20] In February or March I went up by stagecoach from New York to Albany, thence to Bridgewater, and so to Utica, to do my work at Bartlett's school. That finished, made a second trip to the northeast part of the State, collecting in botany and mineralogy with Dr. Crawe, extending the tour to St. Lawrence County, where we found fine fluor-spar and great but rough crystals of phosphate of lime, idiocrase, etc. I wrote some account of these for the "American Journal of Science," the earliest of my many contributions to that journal. Returning toward autumn to Bridgewater, I there received a letter from Dr. Torrey, informing me that the prospects of the Medical College were so poor that he could not longer afford to have my services as assistant. Bartlett's school I had resigned from on account of my prospects in New York. And, in fact, the school was then going down, and he [Bartlett] was transferred soon after to Poughkeepsie, where he flourished anew for a time. I was in a rather bad way. But I determined to go to New York, assisted Dr. Torrey as I could, got out the second part of my "North American Gramineæ and Cyperaceæ." I am not sure whether I was in Dr. Torrey's family or not, or for only a part of the winter. But in the spring of 1835, I went up to my father's house for the summer, with some books, among them a copy of De Candolle's "Organographie" and "Théorie Elémentaire." These or at least the former came from Professor Lehmann,[21] of Hamburg, with whom for a year or two I had corresponded and exchanged plants, or received books in exchange for plants. I had made a still earlier exchange with Soleirol, a French army surgeon, who had collected in Corsica. While at home I blocked out and partly wrote my "Elements of Botany." Returned to New York in the autumn; went into cheap lodgings, arranged with Carvill & Company to take my book. I think they gave one hundred and fifty dollars, which was a great sum for me. We got it through the press that winter. John Carey had then come down to New York, and was a great help to me in proof-reading, and the little book was published in April or May, 1836.

I think it was in the autumn of 1836 that the Lyceum of Natural History, New York, having with a great effort erected their hall, on Broadway just below Prince Street, I was appointed curator; had a room for my use, some light pay, proportioned to light duties, and this was my home for a year or two. There I wrote my papers, "Remarks on the Structure and Affinities of the Ceratophyllaceæ" (which dates February 20, 1837),—not a very wise production, and some of the observations are incorrect; also the better paper, really rather good, "Melanthacearum Americæ Septentrionalis Revisio," published in 1837.

Dr. Torrey had planned the "Flora of North America," but had not made much solid progress in it. I, having time on my hands, took hold to work up in a preliminary way some of the earlier orders for his use. This was to pass the time for a while, for in the summer of 1836 I was appointed botanist to a great South Pacific exploring expedition, which met with all manner of delays in fitting out, changes in commanders, etc., until finally, in the spring of 1838, Lieutenant Wilkes was appointed to the command, the number and size of the vessels cut down, and the scientific corps more or less diminished. The assistant botanist, William Rich, an appointment of the Secretary of the Navy, was to be left out. I resigned in his favor, having been about that time appointed professor of natural history in the newly chartered University of Michigan. As I had thus far done fully half the work, Dr. Torrey invited me to be joint author in "Flora of North America." The first part was printed and issued in July, the second in October, 1838, at our joint expense, my share being contributed from the pay I had been receiving while waiting orders as botanist of the exploring expedition.

By this time we had come to see that we did not know enough of the original sources to work up the North American flora properly, and as Dr. Torrey could not get away from home, I was determined to get abroad and consult some of the principal herbaria. On being appointed professor in the University of Michigan, which had as yet no buildings, I made it understood that I must have a year abroad. The trustees of the university in this view gave me, in the autumn of 1838, a year's leave of absence, a salary for that year of fifteen hundred dollars, and put into my hands five thousand dollars with which to lay a foundation for their general library. I sailed early in November, 1838, in the packet-ship Philadelphia, for Liverpool; went direct from Liverpool to Glasgow; was guest of Dr. William J. Hooker till Christmas—his son, Joseph D. Hooker, was then a medical student; went to Arlary, December 26-7, to visit Arnott; stayed till the day after New Year; thence to Edinburgh for two or three days. Greville was the best botanist, but Graham was the professor, Balfour then a young botanist there. Heard old Monro, Wilson (Christopher North), Chalmers, Traill, Charles Bell, etc., lecture. On way south stopped at Melrose and Abbotsford; coach to Newcastle, Durham (over Sunday), and through Manchester, where rail was taken, to Birmingham and London. Took lodgings till some time in March. Dr. Boott was of course my best friend there. But Hooker and Joseph came up to London for a week. Hooker insisted on taking me in hand as of his party, and so I was introduced to all his friends; took me to the Royal Society, etc.; dined one day with Bentham, to whose house I often went, and who gave me a full supply of letters to the botanists on the Continent. I worked a good deal at the British Museum; Robert Brown was very kind to me, and his assistant, J. J. Bennett, very useful, putting me up to all the old collections and how to consult them. At Linnæan Society, thanks to Boott, had every facility for the Linnæan herbarium. Old Lambert too; he had the Hookers and myself at dinner, and gave me as good opportunity as he could to consult the Pursh plants, etc., in his herbarium, which, not long after, was scattered, but it was in his dining-room, which was very much lumbered, and to be reached only at certain hours. Lindley had me down for a day to his house at Turnham Green, and a little dinner at the close. First visited Kew with the Hookers; called on Francis Bauer, who lived in a house near the river; found him at ninety making beautiful microscopic drawings to illustrate the genera of ferns; and Hooker then arranged for their publication in the well-known volume for which he furnished the text. Saw not rarely N. B. Ward, who lived at Wellclose Square in Wapping, and whose cultivation of plants in closed cases attracted much attention. Went with Ward one day to dine with Menzies, then over ninety; he lived, with a housekeeper, at Maida Vale, or somewhere beyond Kensington.

George P. Putnam, of the firm of Wiley & Putnam, was then resident in London, and through him I managed the expenditure of the money placed in my hands for the purchase of books for the University of Michigan, in a manner that proved satisfactory.

There is still in my possession, but not in reach for ready reference, a file of letters which I wrote home to the Torrey family while I was in Europe. If I were to find them and refresh my memory by them, I should make these notes quite too long. I will therefore trust to memory and touch lightly here and there on my Continental journey. I think it was early in March, 1839, that one morning I took passage on a small steamer from London, Bentham coming to see me off, to Calais; thence diligence for Paris. My lodgings, near the Luxembourg, were not far from the house of P. Barker Webb, to whom I had introductions, and who was very useful to me; he owned the herbarium of Desfontaines. At the Jardin des Plantes were old Mirbel, who occupied himself only with vegetable anatomy, Adrien Jussieu, with whom I corresponded as long as he lived, Brongniart, Decaisne, then aide-naturaliste, and Spach, curator of the herbarium. Jussieu had his father's herbarium in his study. Besides Michaux's herbarium at the Jardin des Plantes, I had also to consult, for a few things, the set taken by the actual writer of the "Flora," L. C. Richard. This I found at the house of his son Achille Richard, botanical professor in the Medical School, living in the Medical Botanic Garden, then occupying a piece of the Luxembourg grounds. The other French botanists I recall were Dr. Montagne, the cryptogamist, a pleasant man, Gaudichaud, whom I saw little of, Auguste St. Hilaire, who I think spent only the winter in Paris. I had an introduction to Benjamin Delessert, who lived in fine style in a hotel in the Rue Montmartre. Lasègue, the librarian, acted as curator to the herbarium (Guillemin had died not long before), which I found occasion to consult only once. I should not forget Jacques Gay, with his large herbarium very rich in European plants. I never dreamed then that so many of them would find their way into our own herbarium. He lived close to the Luxembourg Palace, then the palace of the House of Peers. Gay was the secretary of the Marquis de Semonville, who was a high official there, and so lived near by. He held a weekly reception for botanists, etc., and was a good soul. It was at the herbarium of the Jardin des Plantes that I first made the acquaintance of a botanist of about my own age, Edmond Boissier of Geneva, who was studying some of the plants of his collections in Granada and other parts of Spain, soon after brought out in his work on the "Flora of Granada," etc.

I left Paris in early spring, by malle-poste to Lyons; passed a day with Seringe; steamer to Avignon, diligence to Nîmes, and thence to Montpellier, where I passed two or three days. Delile and Dunal were the professors; saw Bentham's mother and sister, then resident there. Diligence to

Marseilles, steamer to Genoa, Leghorn, a day at each; to Civita Vecchia; a carriage to Rome, along with an English clergyman; thence back same way to Leghorn, Pisa, Florence. Vetturino to Bologna, Ferrara, Padua (Visiani at the garden), Venice; then steamer to Trieste; a day with Biasoletto, including a botanical excursion, and Tommasini. Fell in there with a young artist of New York, whose name I have forgotten. We took places in the malle-poste together to Vienna, but went on two days ahead to Adelsberg; visited the grotto on a fête day when it was all lighted, and all the country people there in gala trim; that night went on by malle-poste. At Vienna, Endlicher, and his assistant Fenzl, but the latter laid up with lame knee. Never saw him afterward, but we had a long correspondence. Steamer up the Danube to Linz, tramway, etc., to the Gmunden See, and so to Ischl; climbed the Zeimitz, all alone, picked my first Alpine flowers; traveled over night to Salzburg, then to Munich; fine times with Martius and Zuccarini, joined the celebration out in the country of Linnæus' birthday,—but not the 24th May; I think two or three weeks later. From Munich to Lindau on Lake Constance; thence to Zurich; up the lake to Horgen; walked over to Art; walked up the Rigi; descended the Rigi to take the boat up the lake, missed it, got a man to put me across in Canton Unterwalden; walked to Stanz, slept, walked next morning to Engelberg, and then over the [Joch?] Pass, and down to Meyringen; next day to Interlaken and the Staubbach, next over the Wengern Alp to Grindelwald, next over the Grand Scheideck to valley of Hassli, up to the Grimsel, passed a Sunday in the snow; walked down to the Rhone glacier and down to Brieg; thence partly on foot, partly char-à-banc, to Martigny; made excursion to the Col de Balme to get a good view of Mont Blanc; back to Martigny, down to Villeneuve, and steamer to Geneva. I reached there, I think, July 4; worked there ten days or so, very sharp; De Candolle, father and son, and Reuter[22] the curator; saw again Boissier. Leaving boat at Lausanne, diligence to Freiburg, Berne, Bâle. Got across country, I hardly remember how, to Tübingen, Stuttgart, Heidelberg, Frankfort; thence to Leipzig; made excursion to Dresden, then to Halle, where was Schlechtendal, and where I looked over old Schkuhr's originals of his Carex plates; thence through Wittenberg to Potsdam and Berlin; worked diligently a week in herbarium. Willdenow, Klotzsch the curator; saw old Link, Kunth, and Ehrenberg. Diligence to Hamburg, where was Lehmann, one of my very earliest correspondents. Steamer from Hamburg to London, late in September. Toward the middle of October went to Portsmouth, and came back to New York in a London packet-ship. Steamers were then only just beginning to make regular trips.

Returning, Michigan University was quite ready to give me a furlough of a year or two, without pay; took hold sharp of "Flora of North America," and in beginning of next summer (June, 1840) we issued the parts 3 and 4 of vol. i. Then went at the "Compositæ;" was interrupted a while in summer of 1841, when I went with John Carey, and James Constable for a part of the time, on a botanical trip up the Valley of Virginia to the mountains of North Carolina, getting as far as to Grandfather and Roan.

It was, I think, in the spring of 1841 that the first part of the "Compositæ" was published, *i. e.*, vol. ii. pp. 1-184; the second part, to p. 400, was out the next spring. Sometime in January, 1842, I made a visit of two or three days to B. D. Greene in Boston; the first time I ever saw Boston. Came out one day to Cambridge, dined with his father-in-law, President Quincy; the company to meet us was Professor Channing[23] and Professor Treadwell.[24] Sometime in April, I received a letter from President Quincy, telling me that the Corporation of the university would elect me Fisher professor of natural history if I would beforehand signify my acceptance. The endowment then yielded fifteen hundred dollars a year. I was to have a thousand and allow the rest to accumulate for a while. Meanwhile I was to give only a course of botanical lectures, in the second spring term, and look after the Garden. But more work was soon added. I came in July, in the midst of vacation, before Commencement, which was then in September; got lodgings, with room for my then small herbarium, in the house of Deacon Munroe. Went late in September on an excursion to Mount Washington, by way of the Notch, along with Tuckerman, then living at his father's in Boston. Worked away at "Compositæ," and in the winter went to New York and carried the remainder through the press. it was issued in February, 1843.

I must not forget that my little "Elements of Botany" had been sold out, and the publishers, Carvill, had gone out of business or died. I prepared in 1841-42 the first edition of my "Botanical Text-Book;" it was in the course of printing when I was appointed to the Fisher professorship, so that I could put that title on the title-page, and have a text-book for my class.

My first session of college work was over about July 1, 1843. The treasurer, Mr. Samuel Eliot, had given me leave to spend a small sum in replenishing the Botanic Garden. I met my friend and correspondent, William S. Sullivant, who had taken strongly to mosses, early in August, on the Alleghanies beyond Frostburg, Maryland (the railroad went only to Cumberland), he coming from Columbus, Ohio, I from Cambridge. There we bought a span of horses and a strong country wagon, and set out on the mountain expedition, some sketch of which is given in the "American Journal of Science" for January, 1846. (The first journey is more particularly detailed in the "American Journal of Science," xlii., no. 1; 1842?) When Sullivant left me, at Warm Springs on the French Broad, anxious to get home, I was left in a pretty lonely condition.

CHAPTER II.

EARLY UNDERTAKINGS.

1831-1838.

DR. GRAY'S autobiographical fragment closes abruptly, and is valuable chiefly for the glimpse which it gives of his ancestry and his boyhood. He kept no diary, but he carried on a voluminous correspondence, and his letters thus contain a record of his hard-working, eager life. The earliest tell of the struggle for position, his doubts if his loved science could furnish him a maintenance, and his resolution to make any sacrifice if he could devote himself to its study. His wants outside of appliances for scientific investigation were few, and he had a hopeful temper. He said in later life that when he was ready for anything it always came to him, and he never dwelt upon the hardships of his early years; indeed, he forgot them.

After leaving Fairfield Medical College he divided his years between teaching in Bartlett's school in Utica (some of his old pupils still recall his field excursions with his class, and his eager delight in the search after plants), in journeys botanical and mineralogical, and in some shorter and longer stays in New York, where for a good portion of the time he was a member of Dr. John Torrey's family. Dr. Torrey was a keen observer, a lively suggester of new theories and explanations, most eminently truthful in all inquiries, and a devout Christian. Mrs. Torrey was a woman of rare character, refined, of intellectual tastes and cultivation, great independence, extremely benevolent, and with a capacity for government and control. She was devotedly religious, not only for herself and her own household, but for all who could possibly come within her influence. It was a new experience to the country-bred young man, and she saw in him many capabilities of which he was as yet himself unconscious. He always said that in his development he owed much to her in many ways. She criticised and improved his manners, his tastes, his habits, and especially, together with Dr. Torrey, exercised a strong influence on his religious life. His parents and family were conscientious, good and faithful church members. But they were not people who talked much, and indeed had little direct oversight of their son after he was fourteen years old, when he left home. He never returned to the family roof after that for more than a few months at a time, and his youthful surroundings away from home were of very varied influence; some of them, though never vicious, were of a decidedly irreligious character. When he entered the Torrey family, the difference in the life, the contrast in the way of meeting trials and sorrows struck him forcibly, and the religious side of his nature was roused, a serious interest awakened, which from that time on made always a strong and permanent part of his character.

Dr. Torrey saw the ability of the young student, and writing to his friend, Professor Henry, in February, 1835, to see if a place could not be found for him at Princeton, says:—

"I wish we could find a place for my friend Gray in the college.... He has no superior in botany, considering his age, and any subject that he takes up he handles in a masterly manner.... He is an uncommonly fine fellow, and will make a great noise in the scientific world one of these days. It is good policy for the college to secure the services and affections of young men of talent, and let them grow up with the institution.... He would do great credit to the college; and he will be continually publishing. He has just prepared for publication in the Annals of the Lyceum two capital botanical papers.... Gray has a capital herbarium and collection of minerals. He understands most of the branches of natural history well, and in botany he has few superiors."

His friend, Mr. John H. Redfield[25] recalling him in those early days, writes:—

"He had worked with Dr. Torrey in his herbarium in 1834 and in 1835, and in 1834 read his first paper before the Lyceum, a monograph of the North American Rhynchosporæ, which is still the best help we have for the study of that genus. His bachelor quarters were in the upper story of the building, and there he diligently employed the hours not occupied with other duties in studies and dissections, the results of which appeared in several elaborate contributions to the Annals. Dr. Gray's residence in the building and his position as librarian brought him into frequent and pleasant intercourse with the members of the Lyceum, and in this way began my own acquaintance with him. The interest which he always manifested in making easy the openings to the paths of knowledge for the younger men impressed me greatly. In describing his manner I should use neither the terms 'imperious' or 'impetuous,' but enthusiastic eagerness would better express its characteristic. He had even then something of that hesitancy of speech which he sometimes manifested in later years, a hesitancy which seemed to arise from thoughts which crowded faster than words could be found for them, and I associate his manner of speaking then with a slight swing of the head from side to side, which my recollections of his later manner do not recall. In person he was unusually attractive, his face, bright, animated and expressive, lit up by eyes beaming with intellect and kindness."

Dr. Gray began in 1834 his contributions to the "American Journal of Science." His first paper, printed in May, was "A Sketch of the Mineralogy of a Portion of Jefferson and St. Lawrence Counties, N.Y., by J.B. Crawe of Watertown, and A. Gray of Utica, N.Y.,"[26] and from that time until his death he was a constant contributor of original articles, reviews, and notices of all botanists whose deaths occurred within his knowledge, leaving an unfinished necrology on his desk.

In 1835 his first text-book was written, "Elements of Botany," and he returned to the same title for his last text-book in 1887. He spent a summer at his Sauquoit home at work upon it; and he once gave a lively account of the warm and noisy discussions which he held with his friend John Carey over style and expressions when he was reading the proofs in his boarding-house in New York, to the great interest of all within hearing. He admitted that it was one of the best lessons in the art of writing he ever had.

Dr. Gray, writing for the "New York World" an obituary notice of John Carey, on his death in 1880, says of him, after a short sketch of his life:—

"Mr. Carey was a man of marked gifts, accomplishments, and individuality. His name will long be remembered in American botany. There are few of his contemporaries in this country who have done more for it than he, although he took little part in independent publication. His critical knowledge and taste and his keen insight were most useful to me in my earlier days of botanical authorship. He wrote several valuable articles for the journals, and when, in 1848, my 'Manual of Botany' was produced, he contributed to it the two most difficult articles, that on the willows and that on the sedges....

"Being fondly attached to his memory, and almost the last survivor of the notable scientific circle which Mr. Carey adorned, I wish to pay this feeble tribute to the memory of a worthy botanist and a most genial, true-hearted, and good man."

It is to be regretted that Dr. Gray's letters to his old friend are no longer in existence.

His correspondence with Sir William Jackson Hooker, then professor at Glasgow, Scotland, began in 1835.

TO JOHN TORREY.

BRIDGEWATER, ONEIDA COUNTY, N.Y., January 1, 1831.

DEAR SIR,—I received your letter, through Professor Hadley, a few weeks since, and I embrace the earliest opportunity of transmitting a few specimens of those plants of which you wished a further supply. I regret that the state of my herbarium will not admit of my sending as many specimens of each as I could wish or as would be desirable to you. I shall be able to obtain an additional supply of most of them during the ensuing summer, when it will give me pleasure to supply you with those, or any other interesting plants which I may meet with. I send you a few grasses; numbered; also a few mosses, etc. When you have leisure, you will oblige me by sending the names of those numbered, and rectify any errors in those labeled. If you should be desirous of additional specimens, please let me know it, and I will supply you in the course of next summer.

You ask me whether I am desirous of obtaining the plants peculiar to New York, New Jersey, etc., or of European plants. I should be highly gratified by receiving any plants you think proper to send me; and will repay you, so far as in my power, by transmitting specimens of all the interesting plants I discover. I know little of exotic botany, having no foreign specimens. I am particularly attached to the study of the grasses, ferns, etc. If you have any specimens to transmit to me, please leave them with Mr. Franklin Brown, Attorney at Law, Inns of Court, Beekman Street, who will forward them to me by the earliest opportunity.

During the next summer, I intend to visit the western part of this State, also Ohio and Michigan. I shall devote a large portion of my time to the collection of the plants of the places I visit. If you know of any interesting localities, or where any interesting plants could be procured, please inform me, and I will endeavor to obtain them for you.

Respectfully yours,
ASA GRAY.

BRIDGEWATER, April 6, 1832.

Having a convenient opportunity of sending to you, I improve it to acknowledge the receipt of your letter of October 6, and of the very interesting and valuable package of plants which was duly received a few weeks afterwards. In the course of the ensuing summer, I shall be able to supply you with an additional supply of most of the plants mentioned in your list. Many of these were collected during an excursion to the western part of the State, and are not found in this section of the country.

I have given a copy of this list to my friend Dr. N. W. Folwell of Seneca County, an industrious collector, who is situated in a section rich in plants, and requested him to transmit specimens of these and other interesting plants to you. I think he will be able to furnish you with many interesting plants from that section of country, and I shall be grateful for any favors you may have in your power to confer upon him. I shall be engaged the ensuing summer at Fairfield and at Salina, where I hope to make some interesting collections in natural history. If it is not too much trouble and the specimen is within your reach, may I ask further information with regard to No. 34, in my last package to you. It is a Carex, from the shore of Lake Erie,—growing with C. lupulina but flowering later. Is it not a var. of C. lupulina? from which it appears to differ principally in its pedunculate spikes? It flowers a month later than C. lupulina (August 6).

Will you excuse me for troubling you on another subject? I shall not be able to remain much longer in this place, unless I engage in the practice of medicine under circumstances which will altogether preclude me from paying any further attention to natural history. My friends advise me to spend a few years in a milder climate, our family being predisposed to phthisis, although I am perfectly healthy and robust; and such a course would be very agreeable to me, as I could combine the study of natural history with the professional business which will be necessary for my support. I have thought of the Southern States, but I have for some time been inclined to prefer Mexico, both on account of the salubrity of its climate, and of its botanical and mineralogical riches, which so far as I know have never been very thoroughly explored. My object in troubling you with all this is merely to obtain some information with regard to the natural history of that country. Has the country been explored by any botanist since Humboldt in 1803? And is there still room enough in that branch to repay one for devoting a few years to its investigations?

I am young (twenty-one), without any engagements to confine me to this section of country, and prefer the study of botany to anything else. Although I have not arrived at any positive determination, I have commenced the study of the Spanish language, and find it (with the aid of Latin and French) quite easy. I should be pleased to have your advice on this subject, as you have many sources of information which are beyond my reach. I should be highly gratified if you would state to me what you think of the prospects in Mexico for a person under my circumstances, and whether any other section of country or any other situation presents greater inducements. Under whatever circumstances I may be placed, it will be gratifying to me to continue a correspondence which has, thus far, been so useful to me, and I shall always wish to do all in my power to render it interesting to you. I shall be ready to leave this place by 1st of September next, at which time I shall probably visit New York. Will you write me on this subject as soon as convenient, and very much oblige,

Yours truly,
A. GRAY.

P.S. There is within a circuit of some miles, and at this place, a great variety of fossil organic remains, and I am collecting them as extensively as possible. We find trilobites (Asaphus, and occasionally Calymene), a variety of bivalve and a few univalve shells, etc., both in lime rock and greywacke. The celebrated locality of Trenton Falls you are of course acquainted with. Would a suit of them be acceptable to yourself, or the Lyceum of Natural History, New York? And can they be named, so that I can label my collection from them? There may few of them be of any interest, but if you wish it you shall have a suit containing specimens of all I find.

UTICA, January 2, 1833.

I received your letter of December 25, and have given the subject of which you write a careful consideration. I may say that I have no objection to the situation you propose, if a proper arrangement can be made.

The terms of my engagement here are these. This situation became vacant by the death of Mr. Edgerton in April last. I was recommended by some of my friends, and finally made an arrangement for one year; took charge of a class in botany and mineralogy on 20th May; closed July 30. Have been at liberty until now; have just commenced a chemical course, to continue nine weeks, which will conclude my duties for the year. The compensation is board, room, washing, fuel, and all other expenses of the kind, for the whole year, or as much of the year as I choose to remain here. All expenses of the laboratory are defrayed (which by the way are not likely to be heavy), and in addition I receive $300. The advantages of the situation are, leisure and the means of a comfortable support. The disadvantages, the school is not incorporated and though now flourishing may not continue so, the scholars are too young, the principal wishes to retain too much of the Eatonian plan to suit me, and they have not furnished the means for the chemical course which I had a right to expect. No arrangement has been made for another year, but I have reason to think I shall be requested to remain another year. I am confident my leisure time would be employed to greater advantage if I was situated so as to have access to good libraries and extensive collections.

At present I can be satisfied with a moderate income, sufficient for a comfortable support, for the purchase of a few books, etc.; but that income must be sure; I cannot afford to run any risks about it. I would willingly collect plants the whole summer, take on my hands the whole labor of preparing and arranging them, but as the proceeds would be absolutely necessary for my support, so they should be certain. I am now advantageously situated for the collection of plants, etc., as, if I choose, I can travel every year with a class who will defray my expenses.

If you still desire to make such arrangement, please to state more explicitly the duties you wish me to perform; how much time can be given to collecting plants; what compensation you can afford me, supposing nearly the whole summer is devoted to making collections, and three fourths of the whole to belong to you,—or propose any plan which would be satisfactory to you, and I will let you know, very shortly, whether I will accept it or not. I had rather leave it to yourself than to make any definite proposition at present. I am confident we can make an arrangement which will be mutually beneficial.

I need not say that I wish to hear from you again on this subject as soon as possible, as I must soon make my arrangements for the ensuing season. How large is the class at the Medical College? I have just returned from a visit at Fairfield; they have a class of about 190. In haste,

Yours very respectfully,
A. GRAY.

UTICA, January 23, 1833.

Excuse me for troubling you. I have this day received from Dr. L. C. Beck a sheet of a work, now publishing, entitled a "Flora of the Northern and Middle States," arranged according to the natural system. I have the sheet commencing the species; commences with Ranunculaceæ; it is in 12mo.

As you mentioned that Beck has been very secret in all his proceedings, it occurred to me that very possibly you have heard nothing of it, and I thought it right to let you know. It appears to be after the fashion of De Candolle's "Prodromus," condensed descriptions and fine print. He still keeps his Ranunculus lacustris, and has added a new species to that genus, which he calls R. Clintonii, from Rome, Oneida County, N. Y.; the same as published in fifth edition Eaton's "Manual" under the name of R. prostratus, Lamk. I have never seen their specimens, but have little doubt it is a form of R. repens, which flowers with us from April to September and assumes many forms. Dr. Beck wishes me to send him any undescribed or interesting plants, localities of rare plants, etc. I feel somewhat interested in the work, as I wish it to supersede Eaton's entirely. (I hear Eaton is coming out with a new edition in the spring. I see Beck means to anticipate him.) But all the undescribed plants I have are in your hands, and it would be improper to send him such at present. He has in his hands an imperfect specimen of Nasturtium natans, De Candolle, which I sent him two years ago. He did not know it; supposed it N. palustre, and I do not know whether he has determined it or no. I will tell him what it is. He has that Ophioglossum and probably will publish it. If you please you can publish this, that Scleria, etc., in Silliman, that is, if you think them new. I will send none of these to Beck, but will give him the localities of some of our most interesting plants.

I have not heard from you since I wrote you on the subject of your letter, but hope you will write me soon. If we can make any arrangement for a year, by its expiration you will know whether or not I shall be of any use to you. I wish to be situated in such a manner as will enable me to advance most rapidly in science, in botany especially.

I succeeded, some days ago, in making the chlorochromic acid of Dr. Thomson (of which you spoke to me when at your house), with chromate of lead, instead of bichromate of potash, which I was unable to obtain. It set alcohol, ether, spirits of turpentine, etc., on fire. I did not try it upon phosphorus. Shall prepare it again in a few weeks for class experiments. I am, Sir,

Yours respectfully,

A. GRAY.

UTICA, March 22, 1834.

I thankfully acknowledge the receipt of your letter of the 1st inst., and am delighted to learn that you contemplate giving a course of botanical lectures before you leave the city. I hope the plan will succeed, and that you will have a large and very fashionable class. My journey was as tedious as rain and bad roads could make it. The first night, being alone in the coach, I was upset by the carelessness of a drunken driver. The top of the coach, striking against a stone wall, was broken in; but I escaped, narrowly indeed, without any injury excepting a few rents in my clothes. At the end of the route, I had the satisfaction of seeing the driver dismissed from his employment. On my arrival at Bridgewater I found a child of my friend and former medical preceptor,[27] a favorite little daughter, dangerously, almost hopelessly sick with inflammation of the brain. I was consequently detained several days, and before I left had the satisfaction of seeing the little patient convalescent. I am now in fine working order and busily engaged in my chemical course.

Dr. Hadley called upon me yesterday and I gave him the little "notions" you sent by me. He was much pleased, but was especially delighted with the condensed sulphurous and anhydrous sulphuric acids.

The principal object of this letter is to consult you in regard to some propositions made me by Professor Hadley. Besides his situation in the Medical College, you are aware that he holds the professorship of chemistry and natural science in Hamilton College. He has just concluded his chemical course in that institution, but in the early part of summer he lectures to the senior class upon botany and mineralogy. As they are about to make some alterations in the college building at Fairfield, his presence will be required there, and he wishes me to take his place for the ensuing term at Hamilton College. I ought also to state that Dr. H. accepted that situation with the intention of holding it but a few years, until the college should have surmounted the trouble in which it was (and is) involved, and from which we have pretty good reason to hope, from the exertions now being made, it will soon be extricated, so that the professorships may be properly endowed. He has given notice of his intention to resign about a year hence; by which time, if ever, the college will be able to place several professorships upon a substantial foundation. Dr. H. has expressed to me a strong desire that I should be considered a candidate for the place, and I strongly suspect that to further that object is one reason for his wishing me to act as his substitute during the ensuing summer. My presence there would be necessary from the 1st of June to the middle of July. Dr. H. has been acting under a nominal salary of $500, being engaged there but thirteen or fourteen weeks. For the summer course I should receive $200. Dr. H. insures me $100 immediately, even if he has to advance it himself, and the whole if funds are in the hands of the treasurer; if not, the whole would be received quite certainly within the year. I have only to say further that the college has now one hundred students, is situated in a beautiful village nine miles from Utica, has the best college buildings of any in the State, has a good faculty, etc. I urged the promise I had made of the visit to Georgia, which this plan would entirely frustrate, but promised to give him a definite answer within a fortnight.

I can scarcely think of postponing my southern tour for another season; but the question comes to this, whether, in the present state of my finances, I had better expend $100 in that visit or earn $200 in the same time. I could also, I think, continue my engagements here in July and August, by which a little more of the trash might be pocketed, and return to New York in time to make a September excursion to the dearly beloved pine barrens of New Jersey, and spend the early part of fall in botanical work, and the winter in your laboratory. The term closes here the 23d of April (a little earlier than I supposed); so if the original plan is pursued I shall be in New York by the 26th of that month. If not, I shall be disengaged for a month, a portion of which I should like to devote, with my friend Dr. Crawe, to the minerals of St. Lawrence County. So rests the case. I told Dr. H. that I should write immediately to you, and be governed in a good degree by your answer.

I have such a dislike to the appearance of vacillation which results from changing one's plans when fully formed, that were it not for certain ulterior advantages, and that I wish to comply with the wishes, as far as may be, of a person to whom I am much obliged, I should promptly decline Dr. Hadley's offer.

An idea just this moment strikes me which, in its crude shape, I will communicate. In eight or ten days I can get to the metals. Suppose I could then get excused, and finish my course here next summer in connection with mineralogy, which for these youngsters would do pretty well; reach New York early next month; set out immediately for Georgia, and remain there until the latter part of May; return via Charleston; examine Elliott's herbarium, and return here by the first of June. I may be quite sure that April and May would be healthy, but could there be plants enough collected, especially Gramineæ, to make it an object? Please say what you think of it. If you think it will do, I see no insuperable objection to carrying it into effect.

A few days ago a letter reached me from Professor Lehmann, in answer to my communication eighteen months ago. He is quite desirous of continuing the correspondence. He is now particularly engaged with Hepaticæ, and is anxious to obtain our species, and especially original specimens of those described by the late Mr. Schweinitz, etc. He has sent a box (which by this time I hope has arrived in New York) containing about five hundred species of plants and several botanical books. He also writes that he has applied to Nees von Esenbeck for dried specimens of all the species of Aster cultivated in his garden in order to transmit them with the monograph by that author; but not having arrived in time they will be sent with his next package. I wish to be particularly remembered to Mrs. Torrey and to Mr. Shaw, not forgetting my lively little friends J——, E——, and M——, whom I very much long to see. I had intended long before this to have written to Mr. Shaw, but have not yet had leisure. Please say to him that I am much obliged for the papers he has been so good as to send me. I wish to know whether he has yet apostatized from the anti-tea-drinking society, of which Mr. S. and myself were ("par nobile fratrum") such promising members. Please say to him that I have not yet drunk tea, but am doing penance upon coffee, milk, and water.

May I trouble you for the very earliest possible answer to this, which will much oblige

Yours very respectfully,
A. GRAY.

HAMILTON COLLEGE, June 9, 1834.

Your letter of the 13th ult., with the bundle of books, was in due time received. Yours of the 2d ult. was received at the same time. I can send you no more copies of "Gramineæ,"[28] etc.; all I brought up are subscribed for and delivered. "Major Downing," who subscribes for two copies (one for himself and one for his friend the Gin'ral,[29] I suppose), as well as the other subscribers, must wait until fall. I am lecturing here to a small but quite intelligent Senior class, twenty-six in number, just enough to fill three sides of a large table, and time passes very pleasantly. The small fund for the support of this institution will, I think, be secured, but the trustees do not act in concert with the faculty, and it is rumored quarrel among themselves, so that, unless some changes are effected in the board, I fear the college will not be sustained. I shall remain here five weeks longer, and then have a short engagement at Utica. I have promised to make a visit to the north in August. I wish very much that I was able to remain there six or seven weeks, to examine with attention the vegetation of the primitive region in St. Lawrence and Franklin counties. I cannot doubt that the mountains and the banks of the large streams of that region would furnish a rich harvest of plants. That range is an extension of one from the far north, which, passing between the Great Lakes and Hudson's Bay, crosses the St. Lawrence at the Thousand Islands, and passes through St. Lawrence, Franklin, and Clinton counties. Consequently many sub-alpine plants, such as Anemone Hudsonica, Trisetum molle, Geum triflorum, etc., are found in this region farther south than elsewhere. The mineralogy of the region, also, needs to be farther explored. The expense of such a tour, divided between Dr. Crawe and myself, traveling in a conveyance of our own, will be comparatively trifling.

I find, however, that further supplies of several New Jersey grasses are absolutely required to enable me to make out the necessary number of suits this fall of the first part of my "Grasses." I see also by the list before me that they (with few exceptions) are in good state as late as the 8th or 10th of September, and that they can all be obtained without proceeding farther south than Tom's River; so that I have no alternative but to hasten back to New York, and make a flying trip to Tom's River (or Howel Works at least) early in September. If you meet with Panicum agrostoides, Poa obtusa Muhl., and Poa eragrostis, I shall be much obliged if you will secure for me the needful quantity of specimens. I am making arrangements for securing the bulbs, tubers, and seeds of the rarer plants for Lehmann. I shall take great pleasure in complying with your desire of securing as many as possible for your little garden. Bulbs and tubers I take up after flowering, and place in dry sand. Can you give some instructions as to the best manner of preserving other perennial roots, such as Asters, etc.? If you will give me the necessary instructions, I promise you to spare no exertions to carry them into effect.

I have nearly finished De Candolle's "Théorie Elémentaire." I have devoured it like a novel. It ought to be translated, that it may be more generally read in this country, where something of the kind is much needed. By the way, as soon as you receive Lindley's new elementary work, I hope you will set about preparing an American edition.

This immediate neighborhood is very poor for botanizing. Excepting Cyperaceæ, it furnishes nothing of interest. I shall soon, however, make more distant excursions, so as to include Oneida Lake and the "pine plains." When I return I shall bring with me a huge bundle of plants, which will show that I have not been idle.

TO HIS FATHER

November 21, 1834.

The class at the Medical College is very small, so that I have no salary here at present. But I have a comfortable and pleasant home, and fine opportunities for pursuing my favorite studies, and for acquiring a reputation that must sooner or later secure me a good place. I have work enough thrown into my hands to support me, with my prudent habits, through the winter. I spend my time entirely at the medical college and at my home here at Dr. Torrey's, and hold little intercourse with any except medical and scientific men. I am writing two scientific articles on a difficult branch of botany for a scientific journal or magazine, which will give me a little notoriety. Dr. Torrey and myself went last month to Philadelphia, where we stayed a week. We spent our time almost entirely in the rooms of the American Philosophical Society, and of the Academy of Science. We met most of the scientific and other learned men, and spent our time very pleasantly. You shall hear from me again before long. It is not probable that I shall be up before next summer.

TO HIS MOTHER

Saturday Morning, February 7, 1835.

I do not know when I shall see you. I shall be up sometime during the spring or summer if I live so long, but perhaps not until July or August. It is very probable that I shall stay in the city the whole time. I wish very much to spend a few weeks in Georgia, early in the spring, but I see that I shall not be able to do so. My time is spent here very profitably, and I am advancing in knowledge as fast as I ought to wish, but I make no money, or scarcely enough to live upon. Just at present I am rather behindhand, but think that by next fall I shall, with ordinary success, be in better circumstances. It is unpleasant to be embarrassed in such matters, for I should like much to be independent, and this with my moderate wishes would require no very large sum, and I have no great desire to be rich.

Tell father I am very glad he has brought home the remainder of those boxes from Utica. The burning down of one of the buildings of the gymnasium has broken up that school entirely, and it probably will not be revived. I knew Mr. Bartlett would fail soon, and that accident has only hastened the time a little. He has been insolvent for some time. There was a very severe fire within a few rods of us last week; five or six dwelling-houses and other buildings were burned to the ground. Although it was so near us we were sitting at tea entirely unconcerned. Everything is done by the fire companies, and people who crowd about fires are only in the way, without doing any good.

Let me hear from you soon, and you will hear from me again in due season. The lectures in the Medical College will be finished in about three weeks, and then I shall be a little more at leisure.

I am very affectionately yours,

A. GRAY.

TO HIS FATHER.

NEW YORK, April 6, 1835.

DEAR FATHER,—I have been waiting for some time to see what my plans for the season would be, expecting as soon as that point was determined to write to you. All my arrangements were upset last fall, and the prospects for daily bread have been rather dark all winter—that is for the present; for the future they look as well as I could expect. It is probable now that I shall remain here during the summer; prosecuting the same studies and pursuits in which I am now engaged, unless something else turns up in the mean time....

Tell mother I have for her a copy of Barnes's "Notes of the Gospels," but I want to read it myself before I send it up. Perhaps I can't spare it until I come up. I think you will all be very much pleased with it. I wish I could also send you his "Notes on the Acts and Romans." Please ask Mr. Rogers, or any of your merchants when they come to New York this spring, to drop a line in the post-office for me, that I may take the opportunity of sending home by them. I wish I could come up this spring, but I see that I shall not be able. Do you take a religious newspaper? Please write to me soon. May the Lord prosper you and keep you all.

Yours truly and affectionately,

A. GRAY.

TO W. J. HOOKER.

NEW YORK, April 4, 1835.

DEAR SIR,—Your kind letter of December 11, with the parcel of books you were so good as to send me, were in due time received, for both of which I beg you to accept my thanks. Perhaps you will do me the favor to accept a copy of the second part of the "North American Gramineæ and Cyperaceæ," being a continuation of my attempt to illustrate our species of these families, the plan of which, I am gratified to learn, meets your approbation. I inclose in the same parcel the loose sheets of an unpublished portion of the third volume of the "Annals of the New York Lyceum of Natural History," comprising an attempt at a monography of the genus Rhynchospora. A more perfect copy, with a copy of the engraving, now in the hands of the artist, will be transmitted to you by the earliest opportunity. I also send a little parcel of mosses, nearly all of which were collected in the interior of the State of New York. May I ask you to look them over at as early an opportunity as may suit your convenience, and to return to me the result of your determinations. I do not venture to think that you will find among them anything of especial interest. I very much regret that I am at the present moment unable to forward to you a half a dozen copies of the work of "Gramineæ and Cyperaceæ," the number you so kindly offer to take charge of. A few species are wanting to complete further suits of the first volume, but these I hope soon to obtain. Not to permit your kind offer to pass wholly unimproved, I hereby transmit to you three copies of vols. 1 and 2 which are at the disposal of any of your botanical friends who may desire to possess the work. If an additional number of copies should be needed they can in a very short time be furnished. With high respect, I remain, dear sir,

Yours truly,
A. GRAY.

To WILLIAM JACKSON HOOKER, LL. D.,
Regius Professor of Botany in the University at Glasgow.

TO JOHN TORREY.

SAUQUOIT, N.Y., July 9, 1835.

I am progressing a little with my rather formidable task; in fact I am making haste quite slowly, and am now discussing the mysteries of exogenous and endogenous stems. I have studied little this week, for I found that close confinement was spoiling my health, so I have been taking quite severe exercise almost constantly, by which I am considerably improved already, although my bones ache prodigiously. I have not yet botanized largely. When at Bridgewater I secured all I could find of the new Carex; also C. chordorhiza, which, by the way, Crawe has found in his region. I hope soon to collect more extensively, but in this vicinity there are no plants of especial interest. I have just now a mania for examining and preserving the roots and fruits of our plants (I make notes of everything in a copy of your "Compendium"), and I hope to bring you a collection in this way which will interest, and perhaps be of some use to you. Fruits and ripe seeds are not often to be obtained, at least in a proper state, in our herbaria. I have been examining our Smilax rotundifolia. It is a regular endogenous shrub, although it sometimes dies nearly to the ground, but always sends out a branch from the uppermost node which survives the winter. It branches just as any endogen would, because the terminal bud is killed; the branches are cylindrical, and increase very little in diameter after their production. A cross-section shows the same structure as the rattan, i.e., the vascular and woody bundles are arranged equally throughout the stem. But a great part of the stem is prostrate beneath the surface, and it may be traced back, alive and dead, for several years' growth. In fact I have not yet succeeded in tracing the stem back to the true root; all I have seen are adventitious roots sent of by the nodes of the stem. This is the only endogenous shrub, I presume, in the Northern States. By the way, the term rhizoma must be used much in descriptive botany, and be extended so as to include all subterranean, nearly horizontal stems, or portions of the stem, which produce roots from any part of their surface and buds from their extremity. It occurs in a great part of herbaceous perennials, and can always in practice be distinguished from the root, although it is still described as root in all the books; witness, Hydrophyllum, Actæa, Caulophyllum, Trillium, Convallaria, and so on to infinity.

I am not yet perfectly satisfied about our Actæas; thus the red-berried one is now perfectly ripe, while the berries of the white one are but half-grown; all the red ones, so far as I have seen, have slender pedicels also, yet the leaves and the rhizomata are exactly alike. By the way, while I was botanizing this afternoon, I met with great quantities of Orchis spectabilis, by far the largest and finest I ever saw; their leaves emulating Habenaria orbiculata. If you care for them in the slightest degree, I will secure a sufficient quantity to fill your garden. O. spectabilis will, while in flower, be a very pretty spectacle....

I remain cordially and truly yours,
A. GRAY.

TO HIS FATHER.

NEW YORK, September 28, 1835.

I suppose I have been a little negligent in waiting so long before I wrote home, but in truth I did not wish to write until I had something certain to say, and even now I have very little. I met Dr. Hadley in Utica just at dusk on the evening of the day you left me there, so I stayed all night there, and went to Fairfield next day. I stayed at Fairfield until Tuesday afternoon, then went to Little Falls, and arrived in Albany just in time for the evening boat next day, and was in New York at breakfast next morning.

Since my return I have been very busy, and on the whole very comfortably situated. I have got back to my class in the Sunday-school; both teachers and scholars have mostly returned, for they all get scattered during the warm months of the summer; and we are now going on very well. On my arrival here I found a very fine package of dried plants collected by my friend the Rev. John Diell, chaplain for American seamen in the Sandwich Islands. I set about them immediately, and it has taken me nearly all my time this month to study them, but I have now finished them. I shall send my notes about them to Professor Hooker of Glasgow, Scotland, that he may, if he pleases, publish them in the "Journal of Botany," of which he is the editor. They are of more interest to the people on that side of the water than to us. I have again sat down to writing upon the work in which I have been engaged all summer, and I do not mean that anything else shall tempt me from it until it is finished, although a nice little parcel of weeds from China, sent by S. Wells Williams[30] (son of Wm. Williams), lies at my elbow. As to my book,[31] I am trying to make a bargain with two publishers; the prospects seem pretty fair, and I shall probably get $300, which is the sum I insist on. I shall have a definite answer in a few days. As to my course and occupation for the winter I can say nothing, for I have not hit upon any certain plan. One thing is pretty certain after thinking over the matter quite seriously, and consulting with Dr. Hadley, who is my firm friend in all these matters; I am determined to persevere for a little while yet before I give up all hopes from science as a pursuit for life. I have now, and expect to have, a great many discouragements, but I shall meet them as well as I can, until it shall seem to be my duty to adopt some other profession for my daily bread. I have several plans before me, some of which you would think rather bold; but I have not yet settled upon any of them. As soon as I take any steps at all I will let you know....

I know little of what is going on in the town. I have not been down into the business part of the city over five or six times since I have been here. When Mr. Rogers comes down, if he will let me know where he stops in season, I will see him. I shall write again to some of you in a very short time. Let me hear soon from some of you, and though I have here little time for writing letters, I will give punctual answers. I remain, with love to mother and all the rest,

Very truly yours,
A. GRAY.

NEW YORK, November 17, 1835.

To-day when I go down town I shall subscribe for the "New York Observer" for you, and pay for a year. The "Observer" and the "Evangelist" are both excellent papers, and I hardly know which to choose. I would send the "Evangelist," did not Mr. Leavitt fill it up too much with anti-slavery. One should if possible read both.

I am now boarding at 286 Bleeker Street, but when you write to me you may direct as before, as I am at Dr. Torrey's a part of almost every day. I have a very comfortable and quiet place, for which I pay $4 per week, and keep a fire besides, which I suppose will startle you a little. I hope to obtain the situation of curator to the Lyceum of Natural History in the spring, when their new building is finished. The duties of the situation will take up only a part of my time. I shall have under my charge the best scientific library and cabinet in the city, a couple of fine rooms to live in, and a salary of about $300. But although I can secure pretty strong influence, the best members of the society offering me the place and wishing me to take it, yet it is not certain that we shall bring it about, so I say nothing about it. I shall let you know whenever any changes offer in my situation.

TO JOHN TORREY.

NEW YORK, July 11, 1836.

DEAR DOCTOR,—Since your departure several memoranda of more or less consequence have accumulated around me, and (having not yet heard from you) I will now communicate them, together with whatever intelligence I think will interest you. To begin with the most important. I have now (5 P.M.) just returned from your house, where I found a parcel for you (received by mail from Philadelphia, postage the mere trifle of $1.14-1/2), with the Hamburg seal, and the handwriting of our old correspondent, Professor Lehmann. Suspecting it to contain advice of packages of plants or books, I took the liberty to open it. I found two diplomas in high Dutch. Shade of Leopoldino-Carolineæ Cæsar. academiæ naturæ curiosorum! Hide your diminished head, and give way to the Königliche Botanische Gesellschaft in Regensburg!—which being interpreted means, I imagine, the Royal Botanical Society of Regensburg. Now I know as little of Regensburg and the Regensburg people who have done us such honor as a certain old lady did of the famous King of Prussia; but I ratherly think it means Ratisbon....

Box of plants and box of bones are here; the plants certainly look the more antediluvian of the two. The specimens are wretched and mostly devoid of interest. The bones will be served up at the Lyceum this evening.... On the same day last week I received a letter from Dewey,[32] and another from Carey, and according to both their accounts they must have been in raptures with each other. Dewey sends love to friend Torrey, and Carey kind regards to Dr. and Mrs. T. Dewey says Carey is rather savage upon species, and where Carey has not given him a favorable opinion upon any, it would amuse you to see how Dewey has detailed them to me, in order if possible to save the poor creatures' lives. Dewey has a good spirit and is altogether a most estimable man, and I am sorry that we have to pull down any of his work. I must write him a few things, that it may not come upon him all at once....

Yours truly,

A. GRAY.

TO W. J. HOOKER.

NEW YORK, April 7, 1836.

DEAR SIR,—I take the opportunity of acknowledging the receipt of your two kind letters, which reached me a few weeks since nearly at the same time, one by the Liverpool packet and the other by the Lady Hannah Ellice. Allow me also to thank you for the trouble you have taken in naming the set of mosses, and especially for the beautiful parcel of British mosses you were so good as to send me, which were truly welcome. All British plants are so, as I have next to none in my herbarium; but nothing could be more acceptable than such a complete and authentic suit of the mosses of your country.

As to the Sandwich Island plants, I hardly know what to say. Supposing they might be of some use to you in connection with other collections, I copied the brief notes I made on studying them very hurriedly indeed, and placed them at your disposal. I did not possess sufficient means for determining them in a satisfactory manner, and fear I have committed errors in many cases. You will doubtless detect these at once, and if, on the whole, you think proper to publish them in the "Companion to the Botanical Magazine," may I ask you to revise the paper, and freely make such corrections and alterations as you think proper. In that case, if you think the notes worthy of publication, I should not object; yet you are equally at liberty to use them in any other way. The parcel contained a specimen of a Composita (from Mouna Kea) which puzzled me extremely, and I was unable to ascertain its genus by Lessing. The anthers are free, or slightly coherent, in all the flowers I examined. Since the parcel was transmitted to you I have seen a specimen of Rhus (from Sandwich Islands) resembling the one in the parcel, except in having pubescent leaves. The latter is therefore improperly characterized, and perhaps will prove to be a well-known species. I shall hope to receive other and more complete specimens from Mr. Diell, and if I am so fortunate will gladly share with so esteemed a correspondent as Dr. Hooker. I hope to send you a parcel by the first opportunity that occurs of sending direct to Glasgow: when I will put up specimens of the mosses you desire, and will send a copy of the "Gramineæ and Cyperaceæ" for the gentleman at Paris who wishes it.

It is so troublesome and expensive to get them bound that I should much prefer, if any of your friends and correspondents should desire them, to send the specimens with labels and loose title-pages, at $4 per volume, each comprising, as you are aware, one hundred species. I may in that way furnish larger and often more perfect or more numerous specimens than in the bound copies. I hope to publish the third (and perhaps also the fourth) volume early next autumn.

Allow me to express my thanks for your kind assistance in various ways, and to say that I shall hereafter (D. V.) prosecute the study of our lovely science with increased zeal. I remain, with sentiments of the highest esteem,

Your much obliged friend,
ASA GRAY.

October 10, 1836.

I also beg your acceptance of a copy of a little elementary botanical work published last spring. I do not expect it to possess any particular interest in your eyes; but in this country, unfortunately, no popular and at the same time scientific elementary treatise has been generally accessible to botanical students, and such a work was so greatly needed that I felt constrained to make the attempt, since no better-qualified person could be induced to undertake the labor.

A letter which Dr. Torrey has just received from Mr. Arnott gives me the information that you have honored my attempt at a monograph of Rhynchospora by commencing the reprinting of it in the "Companion to the Botanical Magazine." I might justly be proud that my first attempt should be thought worthy such notice; but I wish it had been delayed until you could receive the monograph "Cyperaceæ of North America" of Dr. Torrey, in which I had occasion, in the revision of our Rhynchosporæ, to make some important alterations and corrections, as well as to introduce a new species and specify some additional localities. The paper referred to I hope you will receive with this letter.

Except a few extra copies, all the sheets of the monograph "Rhynchosporæ" were destroyed by fire soon after being printed, and when reprinted, about a year since, I added a few observations, notes of additional localities, etc. But owing to a want of careful revision I find there are several errors (several of which are quite material), some of the pen and others of the types. I hope these have been detected and corrected in the course of the reprint. I send herewith the sheets of the paper as published here, with such typographical corrections as now occur to me. Would it not be proper to append a reprint of the revision of Rhynchosporæ in Dr. Torrey's monograph, a copy of which I hope will reach you with the present letter. If the specimens I send please Mr. Webb I shall be glad. It is the last perfect set I have. Please make no remittance, since the sum is too trifling, and moreover I may soon have some favors to ask as to its disposal. Indeed, I know not why I should not state that there is some probability that I may soon visit the islands of the South Pacific Ocean as a botanist, in the exploring expedition now fitting out under the orders of our government. I am anxious to engage in this work, and I suppose may do so if I choose, but I fear that the expedition, which, if well appointed and conducted, may do much for the advancement of the good cause of science, may be so marred by improper appointments as to render it unadvisable for me to be connected with it. I therefore at present can merely throw out the intimation that I may possibly accompany the naval expedition which is expected to sail early in the spring, and to spend two years in the southern portions of the Pacific Ocean. If so I hope to decide the matter in time to procure many needed works, etc., from England and France. I must here close by subscribing myself, with the highest respect,

Your obedient servant,
ASA GRAY.

TO HIS FATHER.

NEW YORK, October 8, 1836.

You may recollect that I intimated to you that there was some probability of my changing my situation before a great while. Matters are now in such a state that it becomes proper to inform you that I shall probably be offered the situation of botanist to the scientific exploring expedition, now fitting out for the South Sea by the United States government. This is to be a large expedition, consisting of a frigate, two brigs, a store-ship, and a schooner; it is to be absent about three years. It will sail possibly in the course of the winter, but very probably not until spring. The scientific corps will consist of several persons, in different departments of science, and the persons who will probably be selected are mostly my personal friends: two of them at least having been recommended at my suggestion. The quarters offered us, and the accommodations, will be ample and complete, and the pay will probably be considerable. We hope to obtain over $2500 per year. Had I room here I would write you further particulars, but this will do for the present. I ask whether, if everything is arranged in a satisfactory manner, you are willing and think it best that I should go. I think it not unlikely that the appointments will be made during the present month. A few days ago I was offered the professorship of chemistry and natural history in the college at Jackson, Louisiana (in the upper part of that State, near the Mississippi River), with a salary of $1500 per year. This I at once declined. I do not like the Southern States.

Yours affectionately,
A. GRAY.

NEW YORK, November 21, 1836.

No appointments are yet made in the scientific corps of the South Sea expedition. The difficulties as to the naval officers are only just settled. There are so many who wish to command that it is impossible to please them all. Captain Jones, the commander, is now in town, and I had the pleasure of seeing him this evening at the Astor hotel. He goes to Boston to-morrow to look after the two brigs fitting out at the navy yard there.

The Secretary of the Navy has written me that when the appointments are made in the scientific corps, the chief naturalists will be called to Washington for a few days, for the distribution of duties among them. If the place for which I ask is given me, it is not unlikely that I may be in Washington early next month. I think you cannot expect E. and myself before about Thanksgiving Day, when if she should have recovered we shall have one reason more than usual for returning thanks to the Author of all good. You did not, it appears, think it a matter of sufficient consequence to say anything about my contemplated voyage; or to offer even an opinion about the matter. Perhaps you thought that, like most people, I only asked advice after I had made up my own mind; and you are not far from correct in this supposition. Still I should have been glad to know that you take some interest in the matter.

As soon as anything is determined upon at headquarters I will let you know....

March 21, 1837.

Since I wrote you last I have been to Washington. I was there at the inauguration and for a few days afterwards. We were not sent for by the Secretary of the Navy, so we had to bear our own traveling expenses, which were not small. When the secretary chooses to convene us, which he seems in no great hurry to do, we shall probably be directed to meet at Philadelphia, or perhaps at New York. There seems to be no doubt but that we shall be here until July.

As they do not choose to advance us any pay yet, money will be very scarce with me for a month or two at least. My engagement at the Lyceum terminated at the close of their year, that is, on the last Monday of last month. So, although I occupy my rooms here until the first of May, I draw no salary.

TO JOHN F. TROWBRIDGE.

NEW YORK, November 9, 1837.

DEAR DOCTOR,—Your letter and that of Mrs. T., dated November 7, reached me this afternoon, to which I hasten to reply, as I have been just on the point of writing you for a week past, but have waited from day to day, in the expectation of being able to afford you more definite information than I could have done. It is this, rather than want of time or inclination, that often causes the delay in writing to my friends. The intelligence which concerns us and interests our friends comes in little by little, day by day. Thus, for instance, the scientific corps were ordered to report here to Commander Jones nearly three weeks ago, and they have been here waiting for a long time, for the secretary had neglected to inform Jones of the fact, and he had come back to his home, and only returned here this week. However, we have now reported and shall take possession of our quarters in a fortnight. They are now undergoing some alterations. We have appointed a caterer, advanced each $120, and our stores will now be soon laid in. The purser of our squadron to-day paid us four months' pay in advance, a very seasonable assistance. My bills having been approved by the government I am now paying them off, and must see to getting all my materials packed up and sent to the vessels, which are now lying at the navy yard, Brooklyn.

This will employ me for a day or two. It is impossible even now to tell you the time of sailing with any certainty. My opinion is that we shall get off about the first or before the 10th of December. It is certain that the ships and stores will not be ready within three weeks, and it would not surprise me, after what I have seen, if we should be kept back longer than you expect. Let us once get to sea and you will not see or hear of so much dilatoriness from us.

November 10. I was prevented from closing my letter last evening by the calling of Professor Henry, who has just returned from a visit of nine months to France and Great Britain. I have been very much engaged all day, and sit down now for a little time, hoping to finish a few letters which have been delayed too long already.

December 5.

I am here yet, and am like to be for a month or so. Commander Jones has been sick for two or three weeks, and I am sorry to say there seems little probability that he will be much better ever. He has a bad cough, and raises blood—is of a consumptive habit. As he has been growing worse, he this morning left for Philadelphia, on his way home. It is thus most probable that we shall have a new commander, and a considerable delay is unavoidable. I think the secretary will be put right this winter by Congress.

Do let me know how Mrs. Trowbridge is. Please send this note to my father, as it is a week or more since I wrote. As soon as anything further is known I will let you know.

Yours very truly,
A. GRAY.

July 18, 1838.

DEAR TRO,—I find, by turning over some books that have been lying on my table, four reviews which certainly ought to have been sent you long ago, but which have been forgotten in my great hurry for the last week or two. I will send them, with this, to-morrow: so look out for them. I have not heard from you since I wrote you a pretty long epistle.

On the 10th instant I tendered my resignation, or rather requested to be left out in the new arrangement. I supposed that it would have been accepted and no words made; but instead Mr. Poinsett sends me word to come on to Washington and have a talk with him, to learn more definitely what their plans, etc., are, and thinks he will be able to remove my present dissatisfaction, and if not says I may have leave to withdraw, but urges me not to insist upon resigning without coming on to Washington. Dana and Couthony are also invited to come on, Pickering being already there. Though this request reaches me in such a form that I cannot claim my traveling expenses, and probably shall not get them (which is just like this nasty administration), yet I suppose I must go on. The only difficulty is that I am afraid they will ply me with such strong reasons as to prevail on me to hold my situation, particularly as their new plan has the advantage of leaving home all the blockheads and taking the best fellows; and moreover some other very promising offers that I had have not been brought to bear very directly; in fact I see that I should get nothing satisfactory from them for a year or two. I intend to set out for Washington to-morrow afternoon. I shall endeavor to make a very short stay, and if I come to any determination there I will try to let you know.

I have scarcely time to write another letter; so please send this up to my father, who has not heard from me in a good while.

Yours very truly,
A. G.

TO HIS FATHER.

NEW YORK, August 6, 1838.

I have resigned my place in the exploring expedition! So that job is got along with. I have been long in a state of uncertainty and perplexity about the matter; but I believe that I have taken the right course. I leave here to-morrow, and am obliged to travel as fast as I can go to Detroit. I shall drop this note on the road somewhere: probably at Utica. I must get as near to Detroit as possible by Saturday evening. I hope to return in the latter part of the month; and intend to make you a visit on my way back.

TO MRS. TORREY.

BATAVIA, GENESEE COUNTY, N. Y.
Friday morning, August 10, 1838.

MY DEAR MRS. TORREY,—The place from which I write is a very pleasant and flourishing country village; the shire-town of Genesee County, forty-four miles from Buffalo and about thirty-four from Rochester. Here is your humble servant and correspondent "laid up for repairs." This is, you may say, my first stopping-place since I left New York, from which place I am distant 418 miles. But I may as well begin at the beginning. I left home, as you remember, on Tuesday evening; breakfasted in Albany, dined at Utica, took stage immediately for Buffalo. We took our supper at Chittenango, which Dr. T. will recollect as the Ultima Thule of our peregrinations in the summer of 1826, and near which place we found the Scolopendrium. Riding all night we were at Auburn (a lovely village) by daybreak, and, passing through Geneva, arrived at Canandaigua in time for dinner. We reached Avon, on the Genesee River, by sunset. Here is a famous sulphur spring; and people crowd the dirty hotels and boarding-houses to drink nasty water. We reached the next considerable village, LeRoy, early in the evening; but our next stage, which brought us to this place, only ten miles, was two and a half hours; so it was about midnight when I arrived here, in a very pitiable plight, so thoroughly exhausted I was obliged to leave the coach and betake myself to rest. I was very unwilling to do this so long as I was able to ride, as, had I continued with the coach, I should have reached Buffalo early in the morning and in time for the steamboat, in which case I could expect to reach Detroit Saturday afternoon, making only four days from New York.

I find myself much better this morning, though weak, and so unstable about the epigastrium that I scarcely dare take any food. I have been debating with myself whether to go on directly to Buffalo to-day, and take the steamboat of to-morrow morning for Cleveland, or some other port in Ohio that I may be able to reach by Saturday evening; or to go from this place directly to Niagara Falls, which I could reach before evening, and remain there until Monday morning. I have pretty nearly decided upon taking the former course, as I shall save some time thereby. But I dread a tedious ride in a stagecoach. In either case I hope to have an opportunity of writing again to-morrow evening.

I met Professor Bailey,[33] of West Point, on board the boat in which I came up the river. He had called the evening previous, when both Dr. Torrey and myself were out. He informed me that the professorship of chemistry, etc., was now established by law on the same footing with the other professorships at West Point, and that the pay of all was increased, so that it is now equivalent to that of a major of cavalry; and more than this: he has been successful in obtaining the place for himself. The stage is nearly ready, and I must hasten. Did the doctor meet Mr. Herrick? I have been thinking that, as they do not know each other, the chance of their meeting at the Astor House is but slight. I must have given both him and yourself no little trouble with my expedition trappings; and if Herrick should conclude to stay at home after all, which is not unlikely, we shall lose our labor. However, tell Dr. T. that I will do as much for him whenever he fits out for an exploring expedition!

CLEVELAND, OHIO, August. 12, 1838,—
the 4th day of my pilgrimage.

Ere this reaches you, a letter which I sent to the post-office in Batavia, New York, will probably have come to hand. The coach called for me before I had finished, and I was obliged to take my portfolio in my hand, and finish, seal, and address the letter in the coach during a moment's delay at the stage-office. I arrived at Buffalo a few minutes after sunset; stopped at a hotel not very munch smaller than the Astor House, with accommodations scarcely inferior. Learning that a boat was to leave for Detroit and the intervening ports that evening at eight o'clock I secured a passage. The internal organization of the Bunker Hill (and I believe the other boats on the lake are not materially different) is rather odd, but very well adapted to answer the purpose for which it is intended. All the boats carry large quantities of freight, and the whole space beneath the main deck is occupied by merchandise, and by the boilers and fuel. The deck is crowded with boxes, bales, and casks, many of which are directed to places in the far West yet so distant that they have hardly commenced their journey. The after part is occupied chiefly by a sort of cabin for deck passengers (equivalent to steerage passengers), in which men, women, and children, Dutch, Irish, Swiss, and Yankee, are promiscuously jumbled. It is infinitely better, however, than the steerage of packet-ships. The bow of the boat is occupied by a different set of passengers, viz., eight or ten horses, destined to draw sundry wagons which now occupy a very conspicuous situation in front of the promenade-deck. You would suppose there was no room left for cabin passengers. On the contrary, their accommodations, though by no means splendid, are really very comfortable and complete. They occupy what in a North River boat forms the promenade-deck, which here extends nearly the whole length of the vessel, has a ladies' saloon entirely separate from the gentlemen's cabin, and three or four private state-rooms for families. The gentlemen's cabin is fitted up with state-rooms with three berths in each, and as there was only a moderate number of passengers I was so fortunate as to secure a whole state-room to myself, where I enjoyed very comfortable rest. When I rose, we were approaching the town of Erie, Pennsylvania. I made an attempt, while we were detained at the wharf, to get on shore to botanize: but time would not permit, and I consoled myself with the comfortable reflection that the dry and sterile gravely banks of the lake were not likely to afford me anything worth the trouble. We had a strong head wind nearly all day, so that our progress was not very rapid: the surface of the lake was covered with white-caps, and the boat pitched so as sadly to disturb the equanimity of a great part of the passengers. Indeed, although I was at no time sick, I found it the most prudent course to pass a large portion of the time in a recumbent position; and I was heartily glad when, a little before sunset, we came in sight of Cleveland. One or two passengers, destined for Detroit, etc., landed to pass the Sabbath here, among whom was Mr. Baldwin of Philadelphia, the machinist, a member of Mr. Barnes' church, a very able and interesting man. We are both at the same hotel, and it being much crowded we occupy rooms which open into each other. I had a little time before night-fall to walk through the city (which will ultimately be a very pleasant place, and is now flourishing, but like most Western towns in a very unfinished state). The people show some signs of civilization: they eat ice-cream, which is sold in many places. I tried the article and found it very good,—nearly the same as what I might just at this moment be enjoying at 30 MacDougal Street, were I now there (as I wish I was), for it is more than probable that the notes of the peripatetic vender are falling upon your ear. Returning to the hotel I consulted the city directory, and read an account of the early settlement of this portion of the State, which is the famous Western Reserve once owned by Connecticut and settled mostly by citizens of that State, who brought with them the heretical doctrines and measures which caused the expulsion of the Western Reserve synod last year. But the evening is advancing, and I must break off; and hoping that the approaching Sabbath may be profitable to both of us and that you may be blessed with comfortable health and strength to enjoy it, I bid you good-night.

Sunday evening.—I attended the First Presbyterian Church this morning, expecting to hear Mr. Aikin, the pastor, formerly of Utica; but, instead, we heard President McGuffey of Cincinnati College, who is quite a celebrated man in this State.

Detroit, Tuesday noon.—I improve the first moment I could secure for the purpose to continue my letter, hoping to fill the sheet in time for the next mail.

On Monday (yesterday) morning I went botanizing, but found absolutely nothing. I kept near the shore of the lake that I might see the first steamboat that came in sight, and one was momently expected. It did not arrive, however, until eleven o'clock, and it was a little after noon before we were under way. The wind was very fresh, and the billows of Lake Erie would not have disgraced the Atlantic. It was, however, in our favor, and we made good progress; but for about two hours we had to run in the trough of the sea, so that the boat pitched and rolled sadly. At sunset we arrived at Sandusky in Ohio. The entrance to the bay is very beautiful. The lake is studded with islands of various sizes, all covered with trees, with here and there a house or a cultivated field upon the larger ones. It was dark before we left; the water was still rough. I went into the cabin and read until it was time to occupy my berth. I am not sure whether I told you that I had lost Bishop Berkeley. I left it behind at Avon, where I was too sick to think about it, but the driver promised me faithfully, for value received, to look it up and send it to the stage-office at Buffalo, where I may find it on my return.

I was roused this morning just at daybreak. We were just at Detroit. I established myself at a hotel, got my breakfast, and sallied forth to survey the town, which is larger than I supposed and most beautifully situated. As soon as I thought your friend, C. W. Whipple,[34] might be at his office I called to pay my respects and deliver the doctor's letter. He was not in; but arrived in a few minutes. He is a good-looking man, but I suspect rather older and a good deal fatter than when you knew him. His black hair has a few silver threads mingled with it, but his countenance is youthful and most thoroughly good-natured. We had some conversation; then went to see Dr. Pitcher, but he was not at home: thence to Dr. Houghton's house, which is entirely occupied as a store-house for the stuff collected in the State survey. It is astonishing what a prodigious quantity of labor Dr. H. and his companions have done and what extensive collections they have made. Dr. H. is not now at home but is expected to-morrow. We went next to the

State-House, but did not find Governor Mason at his office. We looked through the building, at their commencement for a State library, etc., where we met some of the dignitaries of the State. We ascended into the cupola which crowns the building, where we have a most beautiful view of the town and region round about, the roads all diverging from the centre, the noble river, which we could trace from its commencement in Lake St. Clair. The people are evidently very proud of the prospect. By the way, I hear that the doctor's protégé Dr. Fischer has been here, and has gone on to Indiana to astonish the people with his new fashion of blowing up rocks. He has performed wonders in this way between this place and New York. Whipple thinks they will have some place for him next winter. The university branch in this place has a vacation soon, and a public examination is now going on; thither we next directed our steps. I was introduced to the principal, Mr. Fitch, to whom they give a salary of $1500 per annum. I am informed that they employ no teachers or principals in any of the branches without first submitting them to a thorough examination. We stayed until the examination suspended for dinner, when I returned to my room, and here you see me engaged.—Sunset. After dinner Mr. Whipple called for me, and we went to see Governor Mason at his house. We were introduced to his sisters.... They live in a very good house, quite elegantly furnished. We stayed only a few minutes, all going to Whipple's office, where a meeting of the board of regents was appointed to be held. It was known that there would be no quorum, so they adjourned until Thursday, when Mr. Mundy is expected back from New York, and a meeting of consequence will be held. I was introduced to Chancellor Farnsworth (who wrote me from the committee), Major Kearsley, Judge Brooks (Whipple's father-in-law) and others. We all went to the examination, which was, as usual, very stupid, and as it closed we stopped in at the Catholic church—cathedral as it is called—and saw the pictures, of which there are several, some of them valuable. I was struck with a portrait of St. Peter, a stout Paddy-looking fellow with a heavy black beard and mustachios, bare-footed, lugging a pair of keys as large as he could grasp! We expect nearly all hands to go to Ann Arbor on Friday. All speak in glowing terms of the beauty of the location for the university. I had a few minutes' conversation with Whipple as to the plan of buildings, etc., which satisfied me, but I wait for more information before I attempt to write you about the matter.

I am, so far, pleased on the whole with the prospects here, and think they are more promising than I had at first supposed. I must break off again, as I see Governor Mason has come, as he promised, to give me a call. I had hoped to conclude and fill the sheet ere this. I find that we had the fortune to come through the lake in rather slow vessels. There are several upon the lake which make the trip between Buffalo and Detroit in twenty-six or twenty-seven hours. These are large and really splendid boats, carrying little freight, with richly furnished cabins. I will try to arrange matters so as to come down one of these boats. To-morrow I hope to botanize a little.... Mr. Whipple has also asked me to take a ride up to the foot of St. Clair Lake. Now I have nearly filled this very large sheet, and it is so dark I can hardly see to finish. I shall look at the office to-morrow for a letter from home.

I was asked to-day if I would stay here until toward winter! I said I had rather on the whole be excused!

How are the girls? I must write to them specially as soon as I can. Does the doctor go regularly to market every morning? I hope to get away from here early next week. Best remembrances to the doctor. Adieu.

DETROIT, August 16, 1838.

My last letter left here, I suppose, in yesterday morning's boat, and will reach New York in four days. Since its last date nothing whatever has transpired here of any interest. Dr. Houghton arrived here yesterday morning, and as it was a rainy day I spent near the whole time at his house. He is a very energetic little fellow, and the account of his adventures in exploring the unsettled portions of the State is very interesting. He has slept in a house not more than a dozen nights since the commencement of his surveys this season. Mr. Whipple was somewhat unwell, and. I saw him but for a few minutes. I am now going round to his office to read the newspapers, as a mail from New York must have arrived this morning.

Thursday evening.—I spent the whole morning with Mr. Whipple, who is really a downright clever fellow in both the English and the Yankee senses of the term. We compared notes fully about the university and everything about the matter we could think of. I obtained all the information he could afford me about what they were doing, and contemplated doing. I told him fully what I wished to do, and in everything I believe we understood each other and agreed wonderfully. This is important, because Whipple, although secretary of the board, is not a member; yet he is the moving spirit of the whole, and throws his whole energy into the work. We owe the plan adopted as to the arrangement of buildings, etc., to him, and he carried it over considerable opposition. As I know it is just what will please the doctor I mention it here. It is to have the professor's houses entirely distinct from both the university building and the dormitories of the students. The grounds are nearly square, and are to be entirely surrounded by an avenue. He proposes to have a university building for lecture-rooms, library, laboratory, etc., but to contain no students and no families; to have two lateral buildings for students and the tutors who have the immediate charge of them. Then to build professors' houses on the other side of the quadrangle, fronting the main building, each with about an acre of land for yard and garden, by which the houses will not only be away from the students, but at sufficient distance from each other to render them retired and quiet. It is quite a point with him that the professors shall have retired, comfortable houses, so that they shall be subject to no annoyance. By the way, Whipple informed me to-day of something that had turned up quite unexpectedly. Your old friend is about to be made a judge. The appointment is expected to be made by the first of next month. He is induced to accept this place because it will release him from the drudgery of professional business and give him nearly six months of leisure each year: which leisure he wishes to devote to the interests of the university. This will make him a member of the board of regents, of which the judges are ex-officio members.

There was to be a meeting of the regents this evening; but as Lieutenant-Governor Mundy had not arrived there was no quorum. It seems that Mundy has not managed well, and has allowed the plans to be delayed, and Davis, instead of sending the plan he promised, is coming out here to see for himself. So it is probable the plans will not all be in for a month or so. Chancellor Farnsworth, the chairman of the committee appointed to confer with me, called to-day, but I was out. I saw him this evening. Whipple had repeated to him the substance of my conversation with him, and I am desired to commit my plans to writing, that he may embody it in his report at the next meeting of the regents. This I am to do to-morrow (D. V.) and to call on the chancellor to-morrow evening, with Whipple, to talk over the matter. There is every reason to believe that my propositions will be adopted. I say nothing about the subject of salary, and avoid the matter's being broached until the rest is settled. I shall leave it for them to propose. If they employ me, according to the plan I shall present, they can't well avoid offering to pay me handsomely. Prospective appointments will be offered erelong (the coming fall or early in winter) to Professor Henry, Professor Torrey, and perhaps one or two others. Whipple expressed a desire to attempt to secure Professor Douglass[35] for the department of engineering, etc. Everything looks well. The board are determined to prescribe a course of studies and training which shall bring the school up at once to the highest standard. I do not think that there exists another board of regents in the country that will compare with this for energy and capability. But I must break off, as I have a pretty important lecture to prepare to-morrow. I am afraid that these long letters, in which I set down everything that happens from morning to night, will prove very tiresome to you; but I have nothing else to write about. I am anxious to get through, when I will return as fast as steamboats and railroads will carry me.

ANN ARBOR, August 20.

I snatch the few moments that are left me ere the arrival of the stage that is to take me to Detroit to complete my journal. I broke off, I think, late on Thursday evening. On Friday I kept close to my room until I had finished my letter to Chancellor Farnsworth. I sallied out about 4 P.M., showed my letter to Whipple, who approved it altogether and insisted upon our calling on the governor and showing it to him, in order that he might drive the committee a little, if it should be necessary. The servant told us his Excellency was not at home, but Whipple insisted upon his looking into his private room, before he was too confident. And there sure enough we found him. Mason will be down erelong to take a wife. With his approval, the letter was sent round to the chancellor. Whipple, Pitcher, Houghton, and myself spent the evening at the chancellor's residence, a very pretty place. Mrs. Farnsworth is very ladylike and agreeable. Both the chancellor and his lady are from Vermont, and are more than usually intelligent. In the morning I started alone for Ann Arbor,—thirty miles by railroad, and ten (the road not being completed) by stagecoach. I left Detroit at nine A.M. (after going to the post office and being much disappointed and grieved to find no letter,—please tell the doctor so), and reached this place about noon. The location is really delightful, and in a very few years it will be the prettiest possible place for a residence. But I must reserve all particulars until I see you, if I am allowed that pleasure; for although there is an attempt to keep me here until after the arrival of Mr. Davis, the architect, who is to be here in about ten days, yet I am anxious, deeply anxious, to get back again. If I wait his arrival I shall necessarily be detained here until about the 10th of September. It would be desirable on many accounts, but—I don't mean to stay.

The grounds for the university are very prettily situated. The only possible fault I can imagine is that they are too level. I have contrived a plan for the arrangement of the grounds which gives satisfaction to the members of the board here, and I think will suit all. I brought letters to Chief Justice Fletcher and Judge Wilkins. I spent the evening at Dr. Denton's, one of the regents, with several gentlemen and ladies, married and unmarried. It having been ascertained that I was unmarried, it was suggested that I might possibly lose my heart; but I assure you I was never in less danger. On Sunday attended the Presbyterian church here. The pastor, an amiable and very pious old man, was to preach his last sermon to-day, the people having grown too wise for their teachers. His morning discourse from the text, "Christ commended his love to us in that while we were yet sinners," etc.,—a very good sermon. In the afternoon his farewell discourse was from Acts xx. 32, and did honor to his heart. (The stage is ready.) At twilight I in fancy transported myself to 30 MacDougal Street, where yourself, the doctor, and the children were singing your evening hymns. I sang to myself, as well as I could, all the hymns you were singing, as I supposed, and wished myself with you. This morning I have been botanizing, and have cured for the doctor some specimens (clusters of Eshcol) of this goodly land. So be prepared for a very favorable report. My pen is abominable, and I have not another moment.

(DETROIT), 8.30, Monday evening, August 20.

A pleasant afternoon ride brought me back again to this place, where my first care was to run to the post office, nothing doubting that I should find a letter; but I was wofully disappointed, and yet it is the 20th of the month! This is too bad. Do beseech the doctor to write; and especially if I should be detained here until the fourth or fifth day of next month, as I fear may be necessary, ask him to write every other day until you hear from me again.

I am glad to get back here again on one account. The fare here, which is no great matter, I assure you, is excellent compared with the hotel at Ann Arbor. Indeed, I have not taken my place at a single dinner-table for ten days without being reminded of Charles Lamb and his memorable essay on Roast Pig. Here he might riot in his favorite dish (which is in my opinion wretched stuff), as one of the aforesaid juvenile quadrupeds, with a sprig of parsley in his mouth, has been regularly presented to my eyes ever since I left the State of New York. I am sadly bothered as to the course I should take. I suppose I might be able to leave here on Thursday of this week, and, staying over Sabbath at Oswego (making no stay at the Falls), arrive at my father's Tuesday evening, and at New York on Friday morning. But before I could reach New York, Mr. Davis, according to his appointment, would be at Detroit, and it is possible that a very few days would enable us to settle almost everything about the arrangement of the grounds, the internal disposition of the university building, and the plan of professors' houses. I feel so strong a hope that the doctor will be persuaded to take a professorship that I have fixed upon the place for his house, should my plan for the arrangement of the grounds be adopted. And I am very desirous to return to you with the plans in my hands, that I may submit them to Dr. T., Prof. Henry, etc., in time to correct our mistakes and suggest improvements. I see also that if I leave now (although I have explained that I made arrangements on leaving to be back by the first of September, and that it is very necessary I should return by that time), I should lose much of the influence I have acquired, and it is more than probable that some error would be committed that we should not see in time to rectify.

I am anxious that the proper means should be adopted to supply the university and houses with water in abundance, and at such a level that it can be taken into the second story of the professors' houses; I think you may imagine one reason why I am so solicitous about this matter. I was pleased to find on my arrival here that this subject had already received much attention, and there is a determination, on the part of nearly all the regents I have conversed with, to effect this object at whatever expense. Of the different plans in contemplation only one, I think, will effectually answer the purpose. I have some hope that the subject will be acted upon at the first meeting after Mr. Davis arrives. Before that time I suspect we shall not be able to secure the quorum necessary for the transaction of this and other matters of business. I hope also to secure an appropriation for the library, and philosophical and chemical apparatus. I feel pretty confident of accomplishing this result by early autumn.

This is my last entire sheet of large paper, so you may expect no more such tedious letters, unless I find more like it. But if I do not hear from you, and that speedily, I shall be very unhappy. Ask Dr. T. to open any letters that may have come from Norfolk or Washington, and apprise me of the contents, or take any steps that become necessary. Adieu, my dear friend. May our Heavenly Father bless and keep you and yours is the sincere prayer of your attached,

A. GRAY.

TO DR. TROWBRIDGE.

NEW YORK, October 1, 1838.

DEAR DOCTOR,—My arrangements are now so far completed that I may say, with as much confidence as we may speak of any event subject to ordinary contingencies, that I hope to sail for London on the first of next month. I am of course hard at work; there is no need to tell you that. The second part of "Flora" we hope, by hard work, to have published about the 20th inst.

Yours truly,
A. GRAY.

TO HIS FATHER.

NEW YORK, November 7, 1838.

I expect to sail to-morrow for Liverpool in the packet-ship Pennsylvania, unless the weather should prove unfavorable, which is not unlikely. The sailing has already been postponed one day, much to my relief, as, although I have not taken off my clothes for two nights, I am not yet quite ready. I hope to get everything in order before I sleep. You can write to me readily at any time.

I have worked very hard for a few weeks past, but I shall now have a fine time to rest. I am in very good health and spirits.

Mrs. Torrey has a fine boy a few weeks old, and is doing well. Kind remembrances to all, in haste,

Good-by,
A. G.

TO HIS MOTHER.

SHIP PENNSYLVANIA, 9th November, 1838.

MY DEAR MOTHER,—These few lines will be sent on shore in a few minutes by the pilot, and will soon reach you. We shall be out of sight of land in less than two hours more, with a fine breeze. The ship has some motion, but I am not at all sick yet. We have a fine ship and every prospect of a speedy voyage. I shall write at once from Liverpool. Good-by again to all. Letters are called for. Good-by; remember me in your prayers.

Your affectionate son,
A. GRAY.

CHAPTER III.

FIRST JOURNEY IN EUROPE.

1838-1839.

IT has been deemed expedient to give a somewhat fuller narrative of Dr. Gray's first visit to Europe than of his subsequent ones. It was then that he formed many personal acquaintances which ripened into lifelong friendships, and received his first impressions of scenes in nature and art which were to become very familiar. His letters home took the form of a very detailed journal, and it is in extracts from this journal, supplemented by letters to other friends, that this narrative consists.

JOURNAL.

ADELPHI HOTEL, LIVERPOOL, 12 M., December 1, 1838.

We came up the Channel with a gentle breeze, and anchored at half-past nine. At ten minutes past ten I set my feet on the soil (or rather the stone) of Old England. We were very fortunate in our ship, having made our voyage in twenty-one days; while the England (in which, you may remember, I once had intended to sail), which left New York on the first of November, came to anchor just ten minutes before us (thirty days). The Garrick, which sailed on the twenty-fifth of October, arrived here only on Saturday. I must close this letter early in the morning....

Evening.—This short English day has been occupied in good part in getting my luggage from the ship and through the custom house. I sallied out a little past nine in the morning; went first of all to a tailor and ordered a coat (which is to be finished and delivered this evening); then dispatched my letters for home by the United States; found our own ship just going into dock (what docks they are! but as we have always plenty of water we do not so much need them in New York); arranged my luggage, and then proceeded all hands to the custom house (a large new building, rather imposing in appearance), where I was detained until past three o'clock. I had fifteen pounds of books to pay duty upon (fifteen shillings), and nothing to complain of as to the manner of the examination.... After dinner, visited the market, which on Saturday evening is full and busy. It is about twice the size of all the New York markets put together, and a sight well worth seeing. I examined everything scrutinizingly, but will not trouble you with my observations....

Sunday evening, December 2.—Went this morning to the chapel of the school for the blind. The chanting and singing was very fine, and the sight an interesting one. But to me the solemnity of the church service is by no means increased by being chanted; heard a tolerable sermon. In the evening heard Dr. Raffles.[36] His chapel is a gloomy structure externally, but very neat and comfortable within. Dr. R. preached the first of a series of discourses "On the most remarkable events in the early history of the Israelites," commencing with the bondage in Egypt, which was the subject this evening; a very good sermon, delivered in an impressive (but rather pompous) manner. I am very anxious to get to Glasgow. I have been living in society, for the last three weeks, by no means to my taste, and most of them are still here. It is not very pleasant to spend a Sabbath alone at a hotel; but I suppose I must needs become accustomed to it.

I was not fully aware, until yesterday, how much cause we had for thankfulness at our safe arrival. The gales which we encountered off the Irish coast have caused a great number of shipwrecks, and it is feared that many lives are lost. The England escaped most narrowly.

Feather's Inn, Chester, Monday evening.—I have, my dear friend, the singular pleasure of writing and addressing to you another leaf of my journal from a city which was founded, according to the directory which lies before me, "in the year, 917 B.C., at which time Jehosaphat and Ahab governed Israel and Judah,"—the only walled and fortified city in England of which the walls are yet in a state of preservation. The city was rebuilt by Julius Cæsar, and was an important Roman station; and there yet remain many vestiges of Roman occupancy; a hypocaust is still to be seen under the hotel in which I am now staying,—so it is said, for I have not yet seen it, having arrived here after dark. But I expect to be very much interested in this queer old town, for which I owe thanks to Dr. Torrey, since it was his recommendation that induced me to come here. I have scampered about the streets this evening, bought some lithographic views, studied the directory, and am prepared for a busy day between Chester and Eaton Hall, should I live till to-morrow. But it is time I should tell you briefly how I got here. This morning soon after breakfast I walked out to the Botanic Garden, delivered a note of introduction to Shepherd,[37] who received me rather politely, inquired after Dr. Torrey, and showed me through the greenhouses. The establishment is not where it was when Dr. T. was here, but was removed further out of town, two or three years ago. The garden occupies eleven acres; the site is well chosen; but being newly planted there is of course little to see. The hothouses are very well, but not extensive; the collections not particularly interesting, except for some old plants that have belonged to the establishment many years.

I took my cloak and umbrella (necessary articles these!), and at 3 P.M. crossed the Mersey in a small uncomfortable black steamboat, about as much inferior to our Hoboken or Brooklyn ferry-boats as a Barnegat wood-schooner is to a packet-ship; and at Birkenhead took an outside seat for Chester (ten miles), though it rained often and blew hard and cold; had a good view of the country until about five miles from Chester, when it grew dark; saw little villages, farm-houses and cottages, cows, etc., all of which is much more interesting to me than the smoky town of Liverpool. I have seen several little things that are new to me. Let us see what I can recollect at the moment. Hedges of holly—those I am pleased with, particularly when sheared and clipped. The prettiest fence is a stone wall over-topped with a close hedge of holly. Ivy in profusion covering great walls, trees, etc., etc.,—we have nothing to compare with it; a flock of rooks,—very like crows, but larger; an English stagecoach,—more of that anon; a coach and four with postilions,—fine. But I must stop here.

P. S.—Liverpool again, Tuesday evening.—I have accomplished a good day's work to-day. Rose early, made the circuit of the city of Chester on the walls before breakfast, explored all about the town; visited the cathedral, walked to Eaton Hall, four miles and back again; and then, finding there was no coach in the morning until nine o'clock, took an evening coach, and returned here ten P.M., much gratified, but a little fatigued; so good-night. A. G.

GLASGOW (WOODSIDE CRESCENT), December 12, 1838.

I do not just now feel like a traveler. I have been for almost a week, if not at home, yet the next thing to it, in the truly hospitable mansion of our good friends here, where I was received with that cordial kindness which you, having experienced before me, can well understand. Indeed I owe it chiefly to you, who I assure you are not forgotten here. Ecce signum. Both Sir William and Lady Hooker call me, oftener than anything else, by the name of Dr. Torrey. I answer to the name promptly, and am much flattered to be your representative.

I have just stuck fast here, busy among the plants from morning till night. I have been out of the house but twice (except to church on Sunday): once a walk into town with Mr. Hooker, Senior (kind and amiable old man, who insists upon taking me about, and showing me whatever he showed you), and once with Sir William to the Botanic Garden. I am anxious to improve every moment here, where there is so much to be done and such ample means. Arnott has written, inviting me to spend some time with him, which I hope to do, visiting him from Edinburgh, there being now no coach to Stirling or Kinross, from Glasgow direct.... Sir William has given me many interesting plants; we have settled many points of interest. He had our new Nuttallia all figured for the Supplement to "Flora Borealis Americana" as a new genus, and we have recently found it among plants from the Snake country, which, with Douglas's and other Californian plants, he is publishing as a supplement to "Beechey's Voyage." I begged him to adopt the name Nuttallia. He offered at once to publish it as of Torrey and Gray, but I would not consent to this, and I am sure you would agree with me. He has in different ways a great share of Nuttall's so far,—Pickeringia for instance (which is a shrubby Baptisia), Kentrophyta, etc. I shall be kept here ten days longer, I think; no one else abroad is so rich in North American botany or takes so much interest in it. I am requested to study all his Sandwich Island plants (including my own parcel here), and make an article for the "Annals of Natural History" while here. I think I will, if on looking over the parcels I think I can do the subject justice. Can't Knieskern[38] safely make the excursion to Sante Fé in the coming spring? If he can, and will work hard, he will make $1000 clear of expenses! All the collectors make money. Hooker is very anxious about it. I hope to find the fifty copies of "Flora" at Wiley & Putnam's on reaching London. I hope you have seen the partner at New York on the subject, and that the "Flora" will be advertised fully in London before I reach there. But I must close. Don't fail to write very often. Sir William and Lady Hooker and all the family, old, young, and middle-aged, all send their most affectionate regards. I sit over against your portrait at dinner. It is very like you....

TO JOHN TORREY.

KINROSS, Wednesday evening, January 2, 1839.

My journal will inform you of all my movements and doings, and also of the arrival of your welcome letter by the Liverpool, while I remained at Sir William's. I am much distressed at the thought of your anticipated engagements with Princeton, and wish very much that you could have felt yourself warranted in delaying until after the expected meeting of the regents of the Michigan university, which was to take place on the 10th of December. While there is the slightest hope remaining I do not like to relinquish the thought that we may hereafter work together and live near each other. The fear that this may not be the case has of late rendered me much more anxious to obtain books and specimens, in order that I may get on by myself in case I shall be compelled to work alone. I need not attempt to tell you how much I have enjoyed my visit to Hooker. He is truly one of Nature's noblemen. We worked very hard for twenty days, and I would have been glad to have stayed as much longer; for as yet I have looked into few books. All the collections of Carex placed in Boott's hands have been returned to Hooker, and I assisted him in arranging them and selecting for his herbarium; in the course of which I have obtained specimens of nearly all the Northern and Oregonian ones, including one or two which have come in recently, of which I have, when there were duplicates, specimens also for you. The return numbers of those sent you were in many cases strangely misplaced, and Boott has often been sadly confounded. He has studied the genus very critically, hypercritically I may say; for he makes new species where we should think there were too many already. We went over Hooker's Grasses in the same way, and I have obtained numerous specimens and much useful information which we shall presently require. On Christmas day Joseph Hooker selected from a large Van Dieman's Land collection a suite of specimens as far as they have been studied (to Calyciforæ), in which there is in almost every instance a specimen for each of us....

In looking over the recent collections from the Snake country, and Douglas's Californian, I recognized a great portion of Nuttall's,[39] but by no means all. There was a single specimen of Kentrophyta in excellent fruit; another of Astrophia, with neither flower or fruit, collected long ago by Scouler and mixed in with a species of Hosackia, to which genus I am not sure that it is not nearly allied. Nuttall has made too many Hosackias! The copy of "Flora," with my notes, has gone round to London, so that I cannot now communicate many curious things noted in the second part. But how did we overlook the Hosackia crassifolia twice over! I am glad you have the fruit of Chapmannia. I am a little afraid of Stylosanthes, of which there is a sort of monograph by Vogel in the current volume of the "Linnæa;" but no plurifoliate ones appear. Hooker has a curious new genus of Chenopodiaceæ, from the Rocky Mountains, figured for the "Icones," which he wishes to call Grayia! I am quite content with a Pig-weed; and this is a very queer one.

At Glasgow, although my stay was prolonged to twenty days, I was unable in that time to accomplish all I wished with Hooker; and you may be sure we lost no time, and that I could spare very little to visit those objects of interest passing by. I did not omit, however, as you may well suppose, to visit the High Church (the old Cathedral), where I spent an interesting hour, having contrived to go there alone that I might enjoy myself in my own way. From this I visited the new cemetery, which occupies the summit of a hill adjacent to and overlooking the Cathedral. On the very summit, raised on a tall column, is a colossal figure of old John Knox in the attitude of preaching, but ever and anon he seems to cast a scowling look down upon the Cathedral, as if he were inclined to make another attempt to demolish its walls. And well he might, for if what I hear be true, I fancy he would find the preaching now heard within its walls almost as destitute of savor as when the shrine of the Virgin Mary occupied its place in the chapel which bears her name. The Cathedral is now undergoing some repairs; the seats, etc., for the church which occupied the nave are taken away, so that the fine nave presents nearly the original appearance. But the crypt, said to be the finest in the kingdom, is now closed and the key in the possession of an architect at Edinburgh, so that I could not obtain admittance. It was in this place, perchance you may recollect, that the first meeting of Rob Roy with Osbaldistone took place. My Scotch reminiscences have been greatly revived to-day. To-day I have for the first time seen and tasted—only tasted—the two Scotch national dishes, viz., singed sheep's head and a haggis!

I had arranged to leave Glasgow on the morning after Christmas, when Sir William insisted on my staying at least over Wednesday to sit for my portrait! I contrived, however, to sit on Tuesday (Christmas day), when I was done in about four hours, in the same style as Sir William's other botanical portraits, and with so much success that it was unanimously proclaimed to be a most striking likeness; in fact the most successful of all the artist's attempts are said to be this and that of Dr. Torrey, by whose side, it seems, I am destined to be suspended!—a compliment with which I may well feel highly gratified. I believe it is a capital likeness.

I dined out only once at Glasgow, at the house of Mr. Davidson, a very rich don who has made all his money in business here.

Late in the day I went into town to secure a place in the early coach for Stirling and also a bed for the night, as well as to select some little Christmas presents for the Misses Hooker. In the evening Sir William had several friends to dinner, and soon after the breaking up of the evening party I took my leave of these kind friends with no small regret; my contemplated visit of ten days has been prolonged to just twice that number. And now, as we have fairly bid adieu to the old year, I must also bid good-by to you for the present, wishing you, not as the mere compliment of the season, but with all my heart and soul,—a happy New Year. The last New Year I well remember; several of its predecessors also I have had the pleasure of spending with you. I pray God we may be preserved and have a happy meeting before another new year comes.

JOURNAL.

KINROSS, Wednesday Evening, January 2, 1839.

I left Glasgow at seven o'clock A.M. on the morning of the 26th December, on the top of a stage-coach bound for Stirling, so famous in song and story,—distant about thirty miles from Glasgow. I arrived about half past ten, in the midst of a heavy rain.

On leaving Stirling for Perth, I took an inside place, as the storm still continued, but it shortly cleared up, and I rode on the outside nearly the whole journey. The only place worth noticing, or rather which I have time to notice, through which we passed was Dumblane, which is just one of those dirty Scotch villages which defy description. If "Jessie the flower of Dumblane" lived in one of these comfortless and wretched hovels I'll warrant her charms are much overpraised in the song. Here I saw for the first time a genuine ruin; that of the large and once important Cathedral, founded in 1142. During the short-lived establishment of Episcopacy in Scotland I think that the good Leighton was for a time rector of Dumblane. Just beyond Dumblane we passed the field of Sheriff-muir, and beyond this, at the little village of Ardoch, I passed, without being aware at the time, the finest and most entire Roman camp in Britain; we passed some fine country-seats on the road; had a long way the distant Grampian Hills, on which "my father fed his flocks," in full view; and somewhat late in a fine moonlight evening, I arrived at Perth. As the stage which passed Arlary left Perth at nine o'clock in the morning, and I could not afford to spend a day here, I of course saw little of this famous town.... A pleasant ride brought me to Arlary at eleven o'clock A.M., and Arnott was by the roadside awaiting my arrival. I was sorry to learn that he is not a general favorite among his brother botanists; but although most of them possess greater advantages, he has but one superior in Great Britain, and in most departments very few equals. He received me with great kindness, and I have spent a few days with him very pleasantly indeed. He is a hearty, good fellow, and improves vastly on acquaintance. I was exceedingly pleased with Mrs. Arnott, who is exceedingly amiable and lively. On Sunday it stormed terribly, so that we were unable to leave the house. On Tuesday I dined with Mr. and Mrs. Arnott, Mr. Wemyss, the clergyman of the parish, another clergyman, etc., at Mr. Barclay's, Arnott's father-in-law, about six miles from Arlary. About one o'clock to-day, taking leave of Mrs. A. I rode with Arnott to Kinross, and leaving Arnott to write some letters at the hotel in the mean time, I took a boat to Loch Leven Castle,—the prison of the lovely and ill-fated Mary Queen of Scots....

On returning to the hotel I found that Arnott had picked up the dominie of his parish, and had our dinner in readiness. The expected coach arrived soon after, but was crowded. I am consequently obliged to wait for the mail which passes about two o'clock in the morning, and by which, if I am so fortunate as to obtain a seat, I may expect to reach Edinburgh before daybreak.

WATERLOO HOTEL, EDINBURGH,
Thursday evening, January. 3, 1839.

This is my first day in Auld Reekie; and my first business, on sitting down by my quiet and comfortable fireside, shall be to give you a brief account of this day's work. After taking a reasonable modicum of tea I spent the whole of last evening at Kinross in writing, until two o'clock, at which hour the mail-coach punctually made its appearance; and there was fortunately room inside. We drew up at the post office at Edinburgh at half past six in the morning (raining as usual). I took possession of a very comfortable, even elegant room, very different from the six feet by nine bedrooms of most hotels. This is the finest hotel I have yet seen; the Adelphi at Liverpool is not to be mentioned in comparison. I threw myself on the bed and slept for an hour or two. On waking I drew up the curtains of my windows, and had all at once a magnificent view of this picturesque city, which startled me. From descriptions and a few prints I have somewhere seen I find I had formed a very correct view of this city, as far as it went. It is the finest town I have seen or expect soon to see. It owes much of its beauty to its peculiar site, and to the manner in which the old town acts as a foil to the new. Immediately after breakfast I sallied forth, walked down the street, uncertain which of my letters of introduction I should first attempt to deliver; decided for Greville;[40] so I crossed the North Bridge, which is thrown not over a river but over a part of the town, into the old town, crossed High Street, passed the huge block of buildings occupied by the university, plain and heavy without, but the spacious court within very imposing; and a few minutes' walk brought me to Dr. Greville's residence, which looks in front upon a large public square, and on the other the green fields extend up almost to the house,—a complete rus in urbe. Dr. Greville received me very kindly, and seemed well pleased to receive Dr. Torrey's letter; made many affectionate inquiries, and urged me to stay with him while I remained in town. I was predetermined to decline all invitations of this kind in Edinburgh, but found I could give no reasons for doing so that would not seem strange. Dr. Greville said he well knew I should be obliged to stay either with him or Dr. Graham,[41] who would never let me off; so, as I thought Dr. Greville would prove the most useful and edifying acquaintance, I accepted his invitation and promised to send my luggage sometime to-morrow. We set out to call on Professor Graham; walked over into the New Town, the squares, rows, terraces, and crescents all very fine; called at Professor G.'s, who was as usual out; left Dr. Torrey's letter and my own card. Left to myself again, after promising to meet Dr. Greville at dinner at the house of a friend of his, I directed my steps to the Castle, which, crowning a high cliff much like that of Stirling, nearly or quite perpendicular except on one side, is visible from almost every part of the city.... Walked far away to Inverleith Terrace to leave my letters for Mr. Nicoll;[42] returned, dressed for dinner, passed an agreeable humdrum evening at a small family party; returned to the hotel, read two American newspapers (little news), found a good fire in my room, and sat down to make these desultory notes. As to all the rest of what I have seen I may have more to say another day. Good-night!

ST. GEORGE'S SQUARE, 12 M., January 4, 1839.

Before I retire to rest I must hastily and very briefly record my doings to-day, just by way of keeping in good habits; as I am engaged to breakfast at an early hour with Dr. Graham I must soon go to bed. Rose at half past nine (recollect I had not slept the previous night),—a snowstorm. Sight-seeing being out of the question, went to the university, just in time to hear the latter part of Dr. Hope's lecture (Light Carburetted Hydrogen and Safety Lamp); fine-studied and rather formal manner,—did not wear his gown or ruffles at the wrist! Experiments few but rather neat. In cutting off flame with wire gauze he varied the experiment in a way I had not previously seen, viz., by throwing a jet of ether upon the gauze, which burnt below but did not kindle above,—a very pretty effect. He looks to be not above sixty-five, although he must be ten years over that age. Next heard Professor Forbes,[43] a handsome man of very elegant appearance; a most elegant and lucid lecturer; delivered my note of introduction from Professor Silliman; received me very kindly, but I was obliged to leave at once to hear a lecture from Professor Wilson, the famous Christopher North, one of the most extraordinary men living, very eccentric, a gifted genius, and a man of the most wonderful versatility of powers. The subject to-day was the Association of Ideas. The lecture was rather striking, original in manner, with a few flights of that peculiar eloquence which you would expect from Christopher North. Next heard Dr. Monro (Anatomy); very prosy; the class behaved shockingly, even for medical students! Lastly I heard Professor Jameson[44] a stiff, ungainly, forbidding-looking man, who gave us the most desperately dull, doleful lecture I ever heard. It was just like a copious table of contents to a book,—just about as interesting as reading a table of contents for an hour would be; I may add just as instructive! Dined in a quiet way with Dr. Pardie, a young physician to whom I brought a letter from James Hogg; his wife is a cousin of James; went from the table to the college to hear a botanical lecture from Professor Graham; returned to tea and spent the evening. I found I had quite unexpectedly met with profitable acquaintance, as Dr. and Mrs. Pardie were active and ardent Christians, of the Baptist persuasion, and people of a very delightful spirit. They were well acquainted with Mr. Cheever of Salem, who spent some time in Edinburgh previous to his journey to Palestine. I passed a very pleasant evening, and promised to call on them again before leaving town. Returned in the midst of a violent snowstorm to Dr. Greville's, where I am now domesticated, having sent up my baggage from the hotel.

Saturday evening.—Rose this morning at half past seven; and at half past eight, according to engagement, went over to the other side of the town with Dr. Greville, to breakfast with Dr. Graham, and then visit the Botanical Garden (deep snow). We looked about the garden, or rather the greenhouses, until afternoon; much gratified with the splendid collections; but the Sabbath draws nigh, and I cannot go on to tell you more about it now. Called on Mr. Nicoll on my return; made a provisional engagement to meet him at breakfast on Monday and examine his sections of woods. Ran about the streets; left a note at the house of Arnott's brother, to make arrangements (as we have done) for visiting Parliament House, etc., on Monday; returned to Greville's, dressed for dinner, and looked over books, etc., until Professor Graham and Dr. Balfour,[45] secretary of the Botanical Society, arrived; dined; passed a pleasant evening; after family worship had a little conversation with Dr. Greville, retired to my room, and now, as I am at the bottom of the page and my watch says ten minutes to twelve,—to bed. Adieu.

Monday evening.—Two days have passed since I have taken up my pen to communicate to you my little diary. I still remain domesticated at Dr. Greville's, where I am received with the greatest kindness, and am as happy as I can be away from home. I like Dr. G. and family much, there is so much true Christian feeling and simplicity. Dr. G. seems much to regret that he was unable to meet Dr. Torrey in Edinburgh. Yesterday was the first Sabbath of the new year, and I heard two sermons adapted to the season; one in the morning, in an Episcopal chapel (the one to which this family belong) from Mr. Drummond, the text being the latter clause of Hebrews viii. 13; a most excellent, faithful, and godly sermon. In the afternoon I occupied a seat Dr. Greville was so kind as to secure for me in the Old Greyfriars (Scotch) Church, which is so crowded that without this precaution you can hardly expect to get into the church when Dr. Guthrie preaches. He is the most striking preacher I ever heard. I could not help comparing him with Whitfield. The text was the first clause of Eccles. ii. 11. I dare not attempt to give you any idea of the discourse. I wish you could have heard it. In this church-yard the remains of the early martyrs of Scotland repose, not far from the Grassmarket, where they were mostly offered up. I stood upon the very spot to-day where they suffered. We had a terrible wind all last night, which, with the rain, carried off nearly all the snow. The morning was so stormy that I could not fulfill my conditional engagement to breakfast with Mr. Nicoll and look at his curiosities. So I repaired to the university at ten; heard Sir Charles Bell,[46] the professor of surgery,—a decent lecturer, but not remarkable. At eleven I heard the celebrated Dr. Chalmers, the professor of divinity. The old man has a heavy, strongly-marked Scotch countenance, which, however, brightens very much when he is engaged in his discourse. His manner is rather inelegant and his dialect broad Scotch and peculiar. But the matter is so rich that he carries all before him. Every word is full of thought, and he occasionally rose to a very powerful eloquence. He is much beloved, and is considered by all parties, perhaps, as the strong man of Scotland. The subject of his lecture this morning was the advantage (and the abuse) of

Scripture criticism. It was a treat to hear him. He paid a high compliment, in the course of his remarks, to our Moses Stuart.

The weather growing by this time more tolerable, I walked about town,—visited the Parliament House, the Library of the Writers to the Signet; passed through the Grassmarket, returned here, looked at plants with Dr. Greville; dined; received a parcel from Sir William Hooker containing a few plants I had accidentally left (a few he had given me). A very kind letter informed me that he would be in London about the same time with me (which I had in part expected, and about which hangs a tale I must write soon), and also a fine parcel of letters of introduction for me, both to persons on the way to London, and also on the Continent,—to Delessert, De Candolle, Martins, Endlicher, Humboldt, etc. Truly he is a kind man; he has laid me under lasting obligations. He asks me to say to Dr. Torrey that his Grace of Bedford is anxious to receive also the Hudsonia ericoides from New Jersey, and he will be greatly obliged if he will send a box of it to Woburn early in the spring. Attended this evening a meeting of the Royal Society, Dr. Abercrombie[47] (author of "Intellectual Powers," etc.) in the chair. Dr. A. is at the head of the profession here; is greatly esteemed, and is a most exemplary Christian. An interesting paper was read by Professor Forbes, of whom I have spoken before; a man whom from his very youthful appearance you could never have imagined as the successful candidate to the professor's chair against Dr. Brewster. But Dr. Brewster is no favorite in Edinburgh. Other distinguished men were there. I was introduced to Professor Christison,[48] had some pleasant conversation; promised, if practicable, to hear him lecture to-morrow at nine A.M., and look at his museum of materia medica. We had tea after the adjournment, according to the usual custom here, which is a very pleasant one. I only count upon two days more in Edinburgh, and have yet much to do. I am anxious to reach London, where I hope there are letters for me. Good-night. May God bless you all, and keep you.

MELROSE, January 10, 1839, Thursday evening.

On the 8th inst., Tuesday, I went immediately after breakfast to the university and heard Professor Christison's lecture, Materia Medica. He is an excellent lecturer. I spent a half hour with him, in looking over his cabinet of preparations, which contains a large number of fruits, etc., preserved in strong brine instead of spirits. I acquired some useful information concerning the best way to close the jars, for which he has some very neat plans. Then I heard Professor Forbes again; elegant as usual, but he did not succeed very well in his experiments. The next hour I had a rich treat. I heard another lecture from Professor Wilson, on the Association of Ideas, which on this occasion he noticed in a more practical view than before. He recited, in his glowing manner, several passages from

Virgil, and a long one from Milton, and gave a long and most eloquent analytic commentary upon each, far exceeding anything of the kind I ever heard before. After visiting the library of the university—a most magnificent room—I set out for Holyrood House.... I bought one or two poor prints, a cast of the seal-ring of Mary, plucked a bit of holly from a bush standing by the place by the altar before which Mary was married to Bothwell, and reluctantly took my leave. There was yet some time remaining, so I set out to climb Arthur's Seat, which rises abruptly behind Salisbury Crags to the height of eight or nine hundred feet. I attained my wish, and had a beautiful view, from the summit, of the city beneath my feet, and the wide country around. I descended more rapidly than I went up, though at some risk to my neck. Returned to Dr. Greville's, where I dined and spent all the evening.

I had engaged yesterday to breakfast with Dr. Graham. I therefore set off early for that purpose; afterward accompanied him to the Garden, examined the grounds, etc., passed some time in the splendid palm-house. I spent some portion of the morning also with Mr. Nicoll, examining with the microscope his beautiful collection of recent and fossil wood in thin slices; learned how to prepare them. Then arranged my affairs to leave Edinburgh in the morning. In the evening Dr. Greville and myself dined with Mr. Wilson (gentleman naturalist), the brother of the gifted Professor Wilson; himself almost equally gifted, but with a more healthy tone of mind. He interested us so much that our stay was prolonged until nearly the "wee short hour ayont the twal," when we parted, after a pressing invitation to visit him at his country residence in case I ever visited Scotland at a more pleasant season. Taking leave of my kind friends the Grevilles, I was early this morning on my way to Melrose. I have been received with the utmost kindness, not only by this agreeable and most excellent family, but among all the acquaintance I have made in Edinburgh. I had purchased for you a collection of hymns, etc., edited by Dr. Greville and his pastor, Mr. Drummond, with which I was very much pleased, and doubt not you would like them much. But Dr. Greville saw it, and afterwards insisted on sending a much handsomer copy to Dr. Torrey, which was accordingly placed in my hands for him. Melrose is about thirty-six miles from Edinburgh, on one of the routes to Newcastle. We came upon the Tweed among a rugged range of hills, at first a very small stream; followed it along the sinuous valley for a long way, until it became a pretty considerable river, for Great Britain; at length the valley grew wider, softer, and in the proper season, doubtless very beautiful. A smaller stream joined it at some distance before us, and as its opening vale came into view, the driver—I beg his pardon, coachman—pointed with his whip to the opposite side and said, "Abbotsford; " and true enough the turrets of this quaint castellated house were distinguishable, in the midst of a grove mostly of Scott's own planting, near the banks of the Yarrow. We soon after crossed the Tweed, at the place where the White Lady frightened the sacristan in "The Monastery; " the scene of which, you know, was laid at Melrose and in the neighborhood. The fine old ruin of Melrose Abbey now came into view, half surrounded by a dirty little Scotch village. Here I abandoned the coach until to-morrow, secured a gig, and was soon on my way to Abbotsford.... I walked back from Abbotsford, noticing more particularly the beauty of the valley, and the fine Eildon Hills which rise behind Melrose, from whose summit, it is said, a very beautiful prospect may be obtained. I then spent the remainder of the afternoon about Melrose Abbey, the most beautiful ruin I have ever seen or expect to see; more beautiful than I had imagined, and just in that state of dilapidation in which it appears to the greatest advantage as a ruin, for were it entire it would be indeed magnificent. I feel now as if I should never care to see another ruin of the kind; and therefore I shall not visit Dryburgh Abbey (where Scott is buried), as I had intended; although I suppose we shall pass by nearly in sight of it to-morrow. I wish I could bring you some sketch or print that would give you some idea of Melrose, but I fear this is impossible. The exquisite carvings in stone, especially, cannot be appreciated until they are seen. It is said (I forget the lines) that Melrose should be seen by moonlight, and this I can well imagine; but this evening there is neither moonlight nor starlight....

DURHAM, Saturday evening, January 12, 1839.

Soon leaving the Tweed we crossed a range of hills, and came down into the fertile Teviotdale, so famous in border story. Again leaving this valley, we wound our way up the Jedwater, a tributary of the Teviot, rising high up in the Cheviot Hills, just on the line between England and Scotland. We passed Jedburgh, a Scotch village of considerable size and importance, dirty and comfortless of course. Here is an old abbey, which I should have been loth to pass by had I not seen Melrose; thence we ascended the Jed for many a weary mile, until we reached its source high among the Cheviot Hills. Our course was literally "over the mountain and over the moor," for after a tedious ascent we crossed the boundary line at an elevation of fifteen hundred feet above the level of the sea. We were by this time thoroughly drenched with mist and rain; the wind forbidding the use of our umbrellas. We immediately commenced our descent, and just at dusk stopped for a hasty dinner at Otterbourne, so famous in the history of the border warfare as the place of the memorable Chevy Chase. It was too dark to see the cross erected to mark the spot where Percy fell. Pass we over the ride from this to Newcastle, as we saw nothing, though we passed near some places of interest,—Chillingham, the residence of the Earl of Tankerville, for example,—and arrived at Newcastle about nine o'clock in the evening. In the morning I delivered notes of introduction from Hooker and Greville to George Wailes, Esq., one of the active members of the Newcastle Natural History Society; visited their fine building and really splendid museum, especially rich in fossil remains and also in the British birds; made arrangements for correspondence and exchange with the Michigan State Survey; was introduced to a botanist or two; visited the castle built by Robert, brother of William the Conqueror, if I recollect aright, which has stood firmly for many a year, and may stand for centuries more, or as long as the world standeth.... Arrived at Durham at eight in the evening. I called almost immediately upon Professor Johnston[49] and delivered Doctor Torrey's letter and parcel, when we recognized each other as fellow-passengers in the coach from Newcastle, he being a Scotch gentleman,—looking very like my friend Couthouy of the exploring expedition,—whom I was far from imagining would prove to be the professor in the Durham University; took my tea and spent the greater part of the evening with him. He told me he was just about to send a parcel to Doctor Torrey by a friend going next week to America. I must embrace this opportunity to send my letters, now forming a somewhat bulky parcel....

Spent Monday with Professor Johnston in his laboratory, witnessing the progress of some analyses of resins, etc., in which he is now much engaged; also went through the old castle, now used for the university; dined with Professor Johnston at four clock; returned to the hotel.... Took my tea with him, and he accompanied me at half past nine to the coach office whence I took coach for Leeds. I have little to say about Durham University, promising as it is in some respects, because they have adopted the monkish system of Oxford and Cambridge to the fullest extent; the professors and tutors except Johnston are all clergymen; the curriculum includes nothing but classics, a little mathematics, and less logic; their professor of natural philosophy never lectures; they give their professor of chemistry, mineralogy, and geology just fifty pounds a year (nothing for his experiments), and require no one to attend his lectures.

But now I must record some painful news, just learned to-day, which has shocked me exceedingly, but which you will have heard of long ere this reaches you; viz., the loss of the noble ship Pennsylvania, the death of Captain Smith, the first and second mate, and some of the passengers, I hardly yet know how many. I had grown much attached to this ship, and thought highly of its officers, who had been kind to me....

LONDON, January 17, 1839, Thursday evening.

This is dated at this modern Babylon, where I arrived about nine o'clock last evening. I stopped at the White Boar, Coventry Street, Piccadilly; had a quiet night's sleep; rose early this morning, and had breakfasted and was on my way to Dr. Boott's[50] (24 Gower Street) before ten o'clock. I found Doctor B. at home; was kindly received and was introduced to his wife, mother, children, and a brother from Boston who is now with him; spent an hour or two with him; heard that Hooker was in town. Though not a public day went to the British Museum; inquired for Brown (Mr. Brown, for he does not like to be called Dr.), and was so fortunate as to find not only the man himself I was so anxious to set my eyes on, but also Hooker, Joseph Hooker, Bennett,[51] and Dr. Richardson.[52] Passed an hour or two. Brown invited Hooker and me to breakfast with him on Saturday morning; went out with Hooker; first to the Linnæan Society; introduced to David Don,[53] a stout Scotchman, and looked through the rooms of the society. Don offered to give me every possible facility in my pursuits, but of course I said nothing to him about Pursh's[54] herbarium at Lambert's, of which he was formerly curator; for since he married Lambert's housekeeper, or cook, I forget which, Lambert will not allow him to come into the house. From here Hooker took me,—stopping by the way at Philip's, one of the most eminent painters, whose gallery we saw,—to the house of Lambert[55] himself, the queerest old mortal I ever set eyes on. But Carey's description of the man was so accurate that I should have known him anywhere. I was of course invited to breakfast with him any morning at nine; he showed us his Cacti stuffed with plaster of paris, among others a very curious one called muff-cactus, which really looks just like a lady's muff and is not much smaller. Lambert's specimens are the only ones known, and he gave for them something like a hundred guineas,—the old goose! A woman has the care of his collections in place of Don. She stuffs the cacti and seems quite as enthusiastic as old Lambert himself. We went next to the Horticultural Society's rooms in Regent Street in hopes to find Mr. Bentham; but instead we met Lindley, who received us very politely; he asked me to send him my address the moment I was settled in lodgings.... Here I parted from Hooker for the present, declining an invitation to join him at the dinner of the Royal Society's Club, for which I was afterwards almost sorry, as I should have met there Hallam, the historian, and some other distinguished men, as also Brown, whose peculiar dry wit is said to have abounded greatly. Hooker seems as anxious to serve me and aid me here in London as at his own home. He is the most noble man I ever knew. Thence I took a cab and drove into the City, through Temple Bar, down Fleet Street; drove round St. Paul's, to the office of Baring Brothers & Company, who are to be my bankers and to whom my letters here may now be addressed; thence to the office of Wiley & Putnam in Paternoster Row; did not see Mr. Wiley, but learned that the copies of our "Flora" had not arrived, which I am very sorry for, and don't know how to account for it; called at C. Rich's, but found no letters, which was a sad disappointment indeed; thence back here to dinner. At eight o'clock went to Somerset House to attend a meeting of the Royal Society, where again I met Hooker and Dr. Richardson. Brown was also present, for the first time in eight years. Royle[56] was in the chair, at which the botanists present sneered much, as they evidently think him too small a man to fill the seat occupied by Newton, etc. I don't know how he happened to be one of the vice-presidents. I was introduced to him after the meeting, as also to many others. J. E. Gray,[57] who was very polite, gave me and Joseph Hooker tickets for Faraday's lecture of to-morrow evening, invited me to dine with him to-morrow, etc. I was glad to make the acquaintance of Mr. Criff[58] (or Clift) the curator of the Hunterian Museum, the man who exposed Sir Everard Home, who invited us to come and see that museum. While we were conversing, a gentleman, whom Hooker did not at the time recognize, addressed us, and after some conversation with me asked me if I would like to be introduced to Sir Astley Cooper, and see his museum. I answered of course that it would be a great gratification, when he introduced himself as Bransby Cooper, the nephew of Sir Astley,—of whom I have heard formerly not a little,—gave me his address, and Joseph Hooker and myself are to call on him on Monday next. I was introduced also to Dr. Roget,[59] but saw not so much of him as I could wish; so you see I have met more distinguished men in one day than I might elsewhere meet with perhaps in a whole life. But I must break off; I am engaged to breakfast in the morning with Hooker, to meet also Dr. Richardson....

WHITE BEAR, PICCADILLY, 18th January, 1839, Friday evening.

I am not yet in private lodgings, but hope to be so to-morrow. You must not expect me to mention half the things I see in a day here in this busy metropolis, where as yet everything I have seen has been viewed in the most desultory manner. I breakfasted with Hooker and Richardson, who left me for a half hour at the Adelaide Gallery, where I saw very many things to interest me, which we will not stop to talk of now, as I hope to be there again; among other things, a live Gymnotus or Electrical Eel, which gives powerful shocks, they say, for I did not choose to feel it myself. Thence we visited the Museum of the Zoölogical Society, for which Dr. Richardson not only procured us free admittance, but procured for us an order to visit the Zoölogical Gardens; made calls with Hooker, whom Joseph and I left with the Chancellor of the Exchequer in Downing Street, while we passed by Westminster Hall and Abbey down to Bentham's, who has a beautiful residence as retired as the country. Found Bentham an exceedingly pleasant and amiable man; spent an hour or two, till Hooker came in; accepted an invitation to dine with him to-morrow; went into the City; introduced to Richard Taylor,[60] at his printing-office; were all invited to breakfast on Tuesday morning next; went to Longman's famous bookstore and warehouse; one of the young Longmans politely showed us over the building, showed us room after room filled with solid literature,—a most surprising quantity; went by St. Paul's again, saw the Bank, etc.; took an omnibus again to West End; passed by the London University, etc. Joe Hooker and I went to dine with J. E. Gray, who has taken it into his head to show us no little attention; he has lately married a rich wife, a widow, much older than himself; I was quite pleased with her. Went to the Botanical Society,—poor concern; and then to hear Faraday give the first lecture of the season at the Royal Institution, Mr. Gray having kindly offered us tickets. I was unexpectedly introduced to Faraday just before the lecture; pleasant man, with a very quick and lively expression of countenance. The lecture was on Electrical Eels, etc.; most elegant lecturer he is; brilliant and rapid experimenter. I hope to hear him again.

Saturday evening, January 19.—I am now in lodgings, No. 36 Northumberland Street, near Northumberland House, Charing Cross, in the room just vacated by Dr. Richardson; sixteen shillings a week, and a shilling for my breakfast when I choose to take it here. It is half past eleven. I have just come in; no fire, but fortunately my occupation for to-day is soon told. Hooker, Joe, and I breakfasted with Brown at his house, and stayed with him until four o'clock in the afternoon! I have a good deal to say about him, but not here. He is a curious man in other things besides botany. He has a few choice paintings, and a few exquisite engravings he has picked up on the Continent. I coveted them for you. They are just what we should be delighted to have. I dressed for dinner, then drove with my luggage to my present lodgings, and then took up Hooker and Joe for Bentham's to dinner at half past six, where we met Lindley and Mr. Brydges; the dinner was just the beau ideal of taste and simple elegance. In the drawing-room coffee was served up, and in a half hour Assam tea. I am greatly pleased with Bentham, and delighted with Mrs. B. But more of this anon. We are to breakfast with him on Monday, and then make up a party to Kew and the Horticultural Gardens. The house he lives in, a pleasant place, plain but tastefully furnished and arranged, was the one where Jeremy Bentham lived....

Tuesday evening, January 22.—I have to account for myself for two days past, but fortunately this can be done in general terms in few words. Were I to enter very fully into particulars I should fill several sheets. Yesterday Sir William Hooker, Joseph, and I breakfasted according to appointment with Bentham, and set out, although the day was rainy, for a visit to the Horticultural Gardens at Chiswick. We went in an omnibus, and I noticed on the way Apsley House (Duke of Wellington), and the monument to his Grace in Hyde Park, near his house (what is the good of honors, indeed, if one cannot see them?), Holland House, which I saw from some distance, etc. We found Lindley at the Gardens, and looked through the grounds. They have very few hothouses as yet, but have just dug the foundation of a very splendid one, which is, however, to form one wing merely of the general plan. We went to Kew, about two miles farther, and looked through those fine old grounds and gardens. The hothouses and the collections in them were much larger and more interesting than I had anticipated. They are particularly rich in New Holland and Cape plants. There is a new conservatory for large plants, a fine one certainly, which cost six thousand pounds, and the roof was taken from the greenhouse at Buckingham Palace, and therefore cost nothing. It seems an extravagant job, and Mr. Bentham feels sure a much better one of the same size could be built for four thousand pounds. While here we paid a visit to Francis Bauer,[61] now eighty-five years old, and much broken down, but still hard at work, and making as beautiful drawings as ever (beyond comparison excellent), and as delicate microscopical examinations. He has lately been working at fossil Infusoria, and showed me figures of Bailey's plate in "Silliman's Journal" which he had copied. He was greatly pleased when I offered to send him specimens of the things themselves. He showed me the original red snow from arctic America, and also his splendid drawings. Returned to town, and dined with Bentham.

This morning we breakfasted with Richard Taylor in the City; and went afterwards to the College of Surgeons, by appointment Hooker had made, to see Professor Owen, and the fine museum of the college under his charge (John Hunter's originally); a magnificent collection it is, in the finest possible order; and the arrangement and plan of the rooms is far, very far better and prettier than any I have seen. I shall make some memoranda about it. We there met Mr. Darwin, the naturalist who accompanied Captain King in the Beagle. I was glad to form the acquaintance of such a profound scientific scholar as Professor Owen,—the best comparative anatomist living, still young, and one of the most mild, gentle, childlike men I ever saw. He gave us a great deal of most interesting information, and showed us personally throughout the whole museum. I am every day under deeper obligations to Sir William Hooker, to whom I owe the gratification of forming so many acquaintances under such favorable circumstances. Hooker stays over night often at his brother-in-law's, Sir Francis Palgrave, the great antiquarian and Saxon scholar, Keeper of the Records, of whom I have read so much in the "British Review." His eldest daughter, Maria, is spending the winter there. On Hooker's return on Monday he was so kind as to bring me an invitation from Lady Palgrave to dine with them on Saturday, which will be the last I shall see of Hooker, as he is to set out on Monday for home. In the afternoon we spent an interesting hour in looking through the vast halls of the British Museum, particularly through the sculpture, the Elgin marbles, Egyptian antiquities, etc. These last are much more grand than I had supposed. Indeed, I was struck with wonder. I hope sometime to spend a day or two in looking through these rich collections. Called on Lyell the geologist.

We dined with Dr. Roget, the secretary of the Royal Society, where we met Sir Francis Staunton, a great Oriental scholar and traveler, Professor Royle, Dr. Boott, and two others whose names I forget. But best of all Dr. Boott brought me a letter from Dr. Torrey, dated December 25 (Christmas), and I soon contrived to get into a quiet corner to read it; right glad I was to hear from home once more; I will answer it to-morrow. We left very early, as Hooker was to go to Hampstead, where Sir Francis Palgrave resides. Joe and I walked with him, till he should find a stage; but as none overtook us and the night was fine we walked the whole way, three or four miles, and having left Sir William safe and sound, and seen Sir Francis Palgrave for a moment, the remainder of the family having retired to rest, Joe and I walked back again to town. I confess I am a little tired, and am quite willing to go to bed. A Dieu.

Wednesday, January 23, 1839.—Breakfasted and dined with Mr. Bentham, and studied plants with him all day and a good portion of the evening, excepting an hour or so in the morning when we walked out, and Bentham took me through the splendid house of the Athenæum Club, and we also visited the National Gallery, and saw fine paintings in great numbers from almost every artist ancient or modern. It is very near my lodgings, and I intend to visit it again. Here are some of West's original pictures, and likewise the paintings or sketches of Hogarth from which his well-known engravings were taken. They are much more expressive than the prints. E. would enjoy many of them very much, and especially some of Wilkie's of the same kind.

I am to take my breakfast in my lodgings to-morrow morning, which I have as yet done but once. I sent yesterday my letter of introduction to William Christy, who lives out of town, and received to-day a most polite invitation to dine with him to-morrow, and meet Hooker and Joe.

Thursday.—Breakfast at home. Call with Joe Hooker on Bransby Cooper, and then on Sir Astley Cooper; pleasantly received, saw some very curious preparations; spent the morning with Bentham, and dined at Mr. Christy's, Clapham Road, where I spent an agreeable evening. Returning, wrote a letter to Dr. Torrey to go by mail to-morrow to Bristol for the Great Western.

Friday evening.—I breakfasted at my lodgings this morning, and afterwards walked out with Sir William and Joe Hooker to Regent's Park; went to the Coliseum to see the Panorama of London, and well worth seeing it is. It will save me a visit to the top of the dome of St. Paul's, I think, for the Panorama is said to be more perfect than nature. I will say no more about it, as Dr. Torrey has seen it. The illusion is perfect, were it not for some unseemly cracks in the sky! We called on Dr. Boott; then went into the City. Our object was to visit the museum at the India House (where the poet Lamb spent so great a portion of his life). I made the acquaintance, of Dr. Horsfield,[62] the curator, who also collected the best part of the museum in Java and India. He is an American, if you can so call a man who has not been in the country since the year 1800. I was much interested with the library, which contains a vast quantity of Indian idols, sculptures, and antiquities, as well as fine Chinese curiosities. It is immensely rich, also, in Indian, Persian, and Arabic manuscripts; the finest in the world in such things. Some of the Persian (Arabic) manuscripts are most beautifully illustrated, or illuminated, and the writing is neater than you can conceive. Here is preserved also an original petition of the India Company to Oliver Cromwell, with the answer in his own rough and strong handwriting.[63] ... We dined at Lambert's, where we found Robert Brown, Mr. Ward,[64] who had been looking for me, and immediately asked me to name a day to see his plants in the Wardian cases, and an evening erelong to examine some thirty or forty first-rate microscopes which he has in his house; also Dr. Bostock, Mr. Benson, a legal gentleman, a great scholar and author; and last, not least, yet certainly almost the last person I should have expected to see, Lady Charlotte Bury (formerly Lady Charlotte Campbell), whom you will remember as the author of that book on the secret history of the court of George IV. and his Queen, of which we read together, that summer, the deeply interesting review by Brougham. Lady Bury is now supposed to be sixty years old, and was for a long time considered as the handsomest woman in Great Britain; she still looks well, though too embonpoint, and dresses like a young lady, with short sleeves. She is of a high family, a sister of the present Duke of Argyll, and is certainly talented; she is said to be quite poor. Her daughters are married into families of rank, except one (Miss Bury) who was with her mother at Lambert's, whom Sir William Hooker thought remarkably handsome, but I did not. As I have not a high respect for Lady Bury's character I did not throw myself into her circle, and saw almost nothing of her the whole evening. We came away early.

Saturday evening.—I paid a visit, this morning, in company with Joe Hooker, to the Zoölogical Gardens in Regent's Park, where we saw all kinds of four-footed beasts, and fowl, and creeping things. There are four giraffes, but none quite so large as those we saw in New York. There were a very fine orangoutang, very gentle and amiable, a curious spider-monkey, and other curious animals in great plenty. The finest residences I have seen in London are those which look upon Regent's Park. Returning, we called upon Lambert, Saturday being a kind of public day with him, and there met that Nestor of botanists, Mr. Menzies,[65] whom I found a most pleasant and kind-hearted old man; he invited me very earnestly to come down and see him, which I will try to do some day. Meanwhile I expect to meet him on Tuesday at Mr. Ward's.

We just had time to go down into the City to call on Mr. Putnam (publisher) and to learn that copies of the "Flora" had arrived, but were not yet cleared from the custom-house; then took the Hampstead coach to dine at Sir Francis Palgrave's. Excepting Hooker and Joe, I almost forget who the guests were. I was not interested in any of them particularly. Sir Francis was very agreeable; his conversational powers are almost equal to his erudition. His lady, who looks very much like Lady Hooker, is, like all that family, learned and accomplished. I was glad also to meet Hooker's eldest daughter.

The boys interested me much; I think I never saw more intelligent lads. Sir Francis asked me to call at the Chapter-House, Westminster Abbey, his office as Keeper of the Records, and he would show me the Domesday Book. How a sight of it would electrify Dr. Barrett! He asked me at dinner the meaning of the term locofoco as applied to a party in the United States. I gave him the story of the meeting in Tammany Hall which gave rise to the designation, which afforded much amusement.

Sunday evening, January 27.—I was better prepared than last Sabbath, for I took pains to call yesterday at the office of the Religious Tract Society, and found where Baptist Noel preached. It is St. John's Chapel, at considerable distance from here. Nevertheless I attended there to-day, and have reason to be glad that I did so, for I heard a most excellent sermon in the morning, from Psalm ciii. 10-12. Mr. Noel is a most simple, winning preacher, and his sermon was the most thoroughly evangelical and earnest I ever heard from an Episcopal pulpit. I wish I could give you some idea of it. I took notes for your benefit as well as I could, and have written them out, but they will give you a very imperfect idea of it. The church, a large one, with double galleries around three sides, was crowded. This afternoon his assistant, Mr. Garwood, preached, and there was room enough, but we had a good sermon. This Mr. Garwood, you may have seen by the papers, has lately been persecuted a little by his bishop, for acting as secretary to the London City Mission. Both he and Mr. Noel are doing much good in raising the standard of piety and active benevolence in the church they belong to. I hope by next Sunday to inquire out Dr. Reed's church. I have not been out this evening, but have employed myself in copying out my poor notes on the morning sermon, which I trust soon to forward to you.

Monday evening, January 28, 1839.—I spent the morning with Bentham, by appointment, with whom I breakfasted and looked at

Leguminosæ until two P.M.; then joined Joe Hooker (took leave of Sir William this morning, who has returned to Glasgow, via Woburn); made calls, among others on Dr. Rostock, who received me very politely; we then dined together at a chop-house; called on Dr. Boott, spent an hour or two in his very pleasant family; then attended a meeting of the Royal Geographical Society, in which all that interested me was a paper by Professor Robinson of New York, on some interesting matters of ancient geography connected with his travels in Asia Minor. The paper was sent to the Geographical Society by a learned German geographer; it excited much interest....

London, January 24, 1839.—I have so far been seeing men and things chiefly, but have had one or two botanical sittings with Bentham, who is a thoroughly kind and good fellow. He immediately had all the remaining parcels of Douglas's Californian and Oregon plants sent down to his house, and has supplied me as well as he could; and a valuable parcel I shall have of them....

I have seen considerable of Brown, and like him much better than I thought, although he is certainly peculiar. The day we breakfasted with him we remained until four P.M., and he offered to show anything I wished at the British Museum. He showed us all Bauer's drawings in his possession (I have since seen Francis Bauer). He has much more general information than I supposed; is full of gossip, and has a great deal of dry wit.

He is growing old fast, and I suspect works very little now, and I fear there is not very much more work now to be expected of him. He knows everything!...

I spent a good part of yesterday with Bentham, and was to have met Hooker at the Geological Society in the evening; but botany prevailed and I stayed with Bentham, and was a little sorry afterwards, as I should have seen at the society Whewell! Daubeny! Chantry the sculptor, etc.—I have bought a colored copy of Wallich's "Plantæ Asiatiecæ Rariores," 3 vols. fol., very fine, for £15; the publishing price was £36,—the present price by Henry Bohn, who has bought up not only this but almost every other expensive British work on natural history, is £26. It is not yet come round from Edinburgh. I will soon send it to you.... I have seen the "Atakta Botanica" of Endlicher, where there is a plate of Ungnadia (not Ungnodia, as spelled in "Companion to the Botanical Magazine"), but no letter-press as yet....

January 30, Wednesday evening.... Yesterday morning Joe Hooker and myself breakfasted together, and then paid a visit to Westminster Abbey, which we examined in every part, from Poets' Corner to Henry VII.'s Chapel....

As we left the Abbey (where, by the way, we were most thoroughly chilled with our long stay), we went into the Chapter House adjoining, a very antique building crammed with old records and musty manuscripts, and Sir Francis Palgrave kindly showed us the famous Domesday Book, which is in a perfect state of preservation; all the writing perfectly distinct, and so plainly executed that we could read it, here and there, with moderate facility. He showed us a copy of a treaty made with France by Cardinal Wolsey, of which the immense seal appended was cut in gold, and of the most elaborate workmanship. We saw also the original papal bull sent to Henry VIII., constituting him "Defender of the Faith"! We went from this to Westminster Hall; saw the large room, which is very fine; looked into the Court of Exchequer, and saw the Lord Chancellor and other judges in their full-bottom wigs, most funny to behold, I assure you; and the barristers with their queer horse-hair wigs, frizzled on the top of their heads, but tied up into nice and regular curls behind, which fall upon their shoulders. The case of the Canadian prisoners was then under consideration. We then rode in an omnibus to the City and visited St. Paul's Church, which, grand as it is, does not show to advantage after Westminster Abbey. The monumental statuary is very fine; some of it I would mention, but the extreme lateness of the hour obliges me discreetly to break off and finish my account of the day hereafter. Bon soir, or rather Bon jour!

Thursday evening.... To commence where I broke off with Tuesday. We went to dine, by appointment, with Mr. Ward, the plant-case man, at three P.M., which hour was appointed for the purpose of showing us the plant-cases, etc., by daylight. Ward is one of the most obliging men I ever knew. I was perhaps a little disappointed in his plants, but this is the very worst season of the year, particularly in London, and his house, which is in the heart of the city, near London Docks, is very badly situated as to light. But I have learned something from him, and feel confident that I shall be able to manage our plant-cases much better hereafter. Menzies was there, and a truly kind-hearted old man he is. I was to have returned in time to spend the evening at Bentham's, but owing to the stormy weather I did not reach my lodgings till it was too late. On Friday (a snowy day) I was out rather late; went to Bentham's, where I spent the whole morning, dined with him and Mrs. Bentham, three in all!—they have no children, and live in the most cosy and quiet way you could imagine—and spent the whole evening with him in labeling plants which he selected for me from his duplicates. To-day, Joseph Hooker having concluded to postpone till this evening his departure for Glasgow, and having written accordingly to Ward to meet us, we visited the famous greenhouses and conservatories of Loddiges. Miss Maria Hooker was with us, having come out from Hampstead for the purpose. It is rather a long ride to Hackney, but we were well repaid. The collection of Orchideæ is immense and very beautiful, but a very small portion is now in flower. The palm-house, ample and magnificent as it is, rather disappointed me; it seemed not so much larger than that of the Edinburgh garden, and the plants are not in such nice order. Loddiges was very kind to me. Ward selected a few pretty plants for Miss Hooker. I forgot for the moment that there was such a world of waters between us, and was on the point of selecting some for you know whom; I am not sure that I did not bring some after all.

Loddiges took us to his house and showed his collection of humming-birds, which is the finest in the world. He had nearly 200 species, and usually several specimens of a kind, very beautifully mounted and arranged. You can't imagine how beautiful they are! They are his great pets, and I do not wonder. I returned through the City, stopped a few moments at the British Museum, dined with Joe Hooker at his hotel near me, and shortly after saw him start for Glasgow. I sent by him a copy of "Outre Mer" to Lady Hooker. At nine P.M. I went to the meeting of the Royal Society, heard a paper read of the Hon. Fox Talbot's on the power of objects not only to sit for, but to draw their own portraits, which has just been making a great noise in France. It is done by the influence of the light of the sun upon paper prepared by nitrate or chloride of silver. Talbot seems to have found out all about it long ago, but the French have published first. I will write the doctor more particularly about it, and send the "Athenæum" containing the account when it appears.

I have neglected to say that I received two days ago a very kind note from Lindley inviting me to come down to his place, dine with him on Sunday next, stay all night, spend Monday at his herbarium, and meet a few botanical friends at dinner, and return next morning. I declined of course the invitation as far as it related to Sunday, but accepted it for Monday, and offered to get down to Turnham Green in time to breakfast with him. This morning I received another note from him, pointing out the way in which I may reach his house in time. I have also a letter from Francis Bauer, inclosing some European Infusoria, in return for a few of Bailey's I gave him. I will send a portion to Professor Bailey.

Friday evening, February 1.—I spent the earliest part of the morning in my own room; then went to Lambert's, and commenced the examination of Pursh's plants. After dining in a simple way by myself, I went to Bentham's, by appointment, to spend the evening in looking out duplicate plants. I found him and Mrs. B. sitting cosily together in the study. We had a cup of tea and some chat, and then fell to work until half past eleven, when I came away walking as usual by Westminster Abbey, of which I often get very good nocturnal views.

Saturday evening, February 2.... Brown has been very kind to me, in his peculiar way. I have seen him but twice since Hooker and I breakfasted with him, but I hope soon to be at work at the British Museum and to see more of him. He is very fond of gossip at his own fireside, and amused us extremely with his dry wit, but in company he is silent and reserved. I have found out also that it does not do to ask him directly any question about plants. He is, as old Menzies told us, the driest pump imaginable. But although he will not bear direct squeezing, yet by coaxing and very careful management any one he has confidence in may get a good deal out of him. He tells me that Petalanthera, Nutt., is a published genus, and promises to give me all the information about it I desire. I asked him some question about the manner in which the vessels of ferns uncoil. He at once remarked, "They unroll like a ribbon"! Quekett has been examining them, so has a botanist in India; all are much interested in them. I placed Bailey's specimens afterwards in his hands and also some of the Infusoria, which he expressed himself much pleased with when I saw him at Lambert's. By the way, the Infusoria were sent by Bailey himself. I delivered also the parcel for Lindley, and gave the rest I had mostly to Dr. Roget, Mr. Lyell,[66] and Francis Bauer, who were all very glad to get them. I have saved a few for Mr. Ward's microscopical party which he is to give on Wednesday of week after next.... I shall also order, for Sullivant, Hooker's "Icones Plantarum," which will be continued, as Hooker furnishes all the matter for nothing and gives the plates, finding paper and everything. Although there is not so much detail as I could wish, yet it is becoming a very valuable collection for a student of natural orders....

Monday evening.—I have seen the original Taxus nucifera, of Thunberg, both leaves and fruit. Arnott should have paid more attention to it. It is very like Torreya! and doubtless a congener,—and so Brown insinuates. I will see more about it soon. A new edition of Lindley's "Introduction to Botany" is preparing! Sullivant wants, I suppose, a microscope of single lenses—a good working instrument—and an achromatic. This last I think I shall procure for him in London, where they produce more perfect instruments than the French. Can you send Bentham the Lindernias? He wishes much to examine them; send good corollas.

Arnott seems to think much more of Nees von Esenbeck than anybody else. It is generally thought he is in his dotage, and a sad, very sad splitter of straws....

I had some thoughts of going to Paris via Leyden, to see if I can coax anything out of Blume, but he seems to have behaved rather strangely to all the English botanists I have yet met with. You ask whom I liked best in Scotland: Hooker is all in all!

A new Antarctic expedition is planned; indeed is settled upon nearly, to be commanded by James Ross. But a part of the administration throw difficulties in the way. If it goes Joseph Hooker is to be the naturalist.... By the way, Corda's "Memoir on Impregnation of Plants" turns out to be mere humbug, and it seems there is little dependence to be placed upon him....

Tell Bailey I am every day getting information that will be valuable to him, in the microscopical way. I have a new correspondent for him, Mr. Edwin J. Quekett,[67] 50 Wellclose Square, London, an excellent microscopist. I will write soon what he wants, and he will send through me some microscopical objects.

P. S.—I have just had the offer of a chance to examine Walter's herbarium as much as I like!—to take it into my possession for a week if I like! and that after I had nearly given up all hopes of it.

February 5, eleven o'clock, evening.... I think I mentioned in those letters how yesterday was spent, viz., that I rose early, took stagecoach for Turnham Green, near Chiswick, where Lindley resides, breakfasted and spent the day. Lindley was certainly very civil. Mrs. Lindley is a quiet lady of plain manners and apparently very domestic habits. Miss Drake, whose name appears as the artist in all of Lindley's plates almost, was present, and is, I judge, a member of his family, and perhaps a relative of Mrs. Lindley. I saw Lindley's splendid "Sertum Orchidaceum," and a much more luxurious work, the "Orchidaceæ of Mexico and Guatemala," by Bateman, a very large-paper work à l'Audubon. We looked over some families together in a desultory way, and I took up the Lupines and compared ours carefully with Lindley's, which were named by Agardh. At dinner met Dr. Quekett and Mr. Miers,[68] a traveler in Brazil. On reaching my room I found a note from Bell, the zoölogist (to whom I brought a letter from John Carey, but left at his house, not being able to see him), inviting me dine as his guest at the Linnæan Club, before the meeting of the Linnæan Society. Fortunately, as I do not like club-dinners, I had previously accepted Bentham's invitation to dine quietly with him and Mrs. B. on that day, so I sent a note of declinature. I have already told you of my failure, by my own carelessness, of seeing the opening of Parliament, which I regret, as I should like to see the peers in official costume, and the peeresses in full dress.

It did not break my heart, but I returned to Bentham's and looked over plants until the hour approached to take my place in the park to see the queen, and—what is finer—her superb horses, with what success I have already said; thence to the Horticultural Society, where I received the welcome letters. After dispatching my parcel of letters I took a cab for Bentham's, as it was raining finely, where we dined in his quiet, elegant way.

I don't think Dr. Torrey saw enough of him, at least in his own house, to appreciate him fully....

You may well infer from my being so much with him that he is my favorite....

Wednesday evening.—After breakfast to-day I went to Lambert's, thinking to finish nearly the examination of Pursh's plants, but I found Lambert on the point of going out, though the morning was unpleasant. So I was obliged to retrace my steps; and as a dernier ressort I went to the British Museum, and commenced my examination of the Banksian Herbarium. Brown was there most of the time, but did very little except to read the newspaper and crack his jokes. I broke off at four o'clock; went down to the City, called on Mr. Putnam, took a parcel of late American newspapers away with me, dined, went up to Dr. Boott's, where I spent the evening so pleasantly that eleven o'clock arrived before I thought of it. It is now twelve. On my return here I found my parcel had arrived from Edinburgh, the beautiful copy of Wallich's work, a very complete and pretty set of British Algæ from Dr. Greville, and some letters of introduction for the Continent which he has obligingly favored me with. I must write a letter of thanks to-morrow....

Went to Ward's to see the tunnel.... We had tea, Miss and Mrs. Ward regaled us with music,—and both play extremely well; then Ward and I looked over plants until nearly half past ten, when we had supper, a very substantial one, and I took my leave, arriving at my lodgings a little after twelve....

Sunday evening, February 10.... This morning I attended one of the larger Methodist chapels, where I heard an excellent sermon from 1 Pet. v. 7: "Casting all your care upon him; for he careth for you." A portion of the Episcopal service was read at the beginning from the desk; but afterwards the clergyman ascended to the pulpit, when the singing and prayers were in the ordinary manner. In the afternoon I went to hear my old favorite Baptist Noel, who was to preach a kind of charity sermon for the infant-schools of St. Clement's, Danes. I felt satisfied that we should have a close and fervent sermon, and truly I was not disappointed.... He preaches ex tempore, but has the most perfect facility of language; the words drop from his mouth without any apparent effort, but he never repeats, and all seems equally important; so unless I could write as fast as he speaks I could give you no proper idea of his discourse. His manner is so exceedingly placid that you wonder how he fixes the attention of his auditors so perfectly. There are many other clergymen who have the same ardent piety, and the number I hope is increasing; so that one cannot help expecting great things from this communion, if it once gets free from the contaminating influence of the political power. These men all preach continually to crowded houses, which is another good sign, and proves that the people are ready to hear sound doctrine. I hoped to have heard another of the same stamp this evening, and went all the way to St. Sepulere's, where Mr. Dale preaches in the evening, but he was out of town....

February 5, evening.—It is not long since I closed a parcel of letters for you, and dispatched them by mail to Liverpool, for the steamship Liverpool, by which I hope they will reach you early. I have since attended a meeting of the Linnæan Society, Mr. Forster in the chair. Lambert never comes now for fear of meeting Don, and also because he is a little piqued, perhaps at not being made president. Brown seldom comes, as he would have to take the chair in Lambert's absence, and he fears he might annoy Lambert, for Brown is extremely tender of other persons' feelings. I was most interested in the nominations to fill up the five vacancies of the foreign associates. They were Carus, Milne-Edwards, Dutrochet, Endlicher, and Torrey. The nomination was signed by Bentham, Brown, Boott, Forster, Owen, etc. I knew nothing of it till just before the meeting, and I may be allowed to say that I felt extremely gratified at such a very handsome compliment paid to my best friend.

Lindley has given me to-day a copy of Griffith's most admirable paper in the last part of the "Transactions Linnæan Society," on the ovula of Santalum, Loranthuns, Viscum, etc., an anatomical paper of the very highest order,—about forty pages, with eleven fine plates. I am going to buy all the other papers on Botany in the Linnæan Transactions which I think valuable. They can be had of Coxhead, who buys sets and pulls them to pieces to sell separately. Let me not forget to tell you that, after having made diligent inquiry of Brown, Bentham, etc., I had nearly given up all hopes of finding Walter's[69] herbarium. I spoke to Lindley yesterday, and he said he knew the son of old Fraser, who would be most apt to know something about it, and would give me his address, by which I could find him if in town. But to-day, just after the adjournment of the Horticultural Society, and while I was glancing over your kind letters, Lindley came to say that he had found Walter's herbarium for me! He introduced me to Mr. Fraser, to whom it belongs, though not immediately in his possession, who offered to send it up for my examination to the Horticultural Society's rooms, or anywhere I chose. I hope to get at it, with Bentham, about Friday. I shall be anxious to let you know the result....

I am most clearly of the opinion that any person who will make extensive collections of North American plants, both Northern and Southern, and include also a good collection from Santa Fé, the Platte country, etc., have his sets named according to our work, and who would devote four or five years to the business, could, if he were really industrious and prudent, realize $1000 per annum (clear). He should continue my grass-book for one thing, giving loose sets only for the present price, and while from time to time he sells off collections as he can, should retain some fifty sets in all the most interesting genera or small families, get all the species, and publish them in monographic sets. Knieskern could make, with the aid we would gladly furnish, at least ten times as much money, as long as he lives, as he ever will at physic, besides being engaged in a much pleasanter way. I know how all this should be managed now. Now for Dr. Clapp. Tell him that Brown informs me that he does not think jewel lenses can be depended upon as possessing any advantage over glass. He has an excellent sapphire one, but that is a mere chance, and no other has been made anything like it. They are now almost never made, and appear to be going wholly out of use. His other matters I will take in hand, but he must not expect $20 to procure a doublet 1/40th inch focus, two micrometer glasses, and a case of dissecting instruments. I have some engagements before me with microscopical people, and when I get from them all the information I can, I will set about these affairs more understandingly....

Saturday evening, February 9.—I have been engaged nearly the whole day upon the herbarium you so much wished to examine, viz., that of Walter. I have not yet finished it, and find the examination very tedious, as the specimens are very often not labeled, except with the genus in his "Flora," so that I have first to make out his own species, and then what they are of succeeding authors.

The specimens are mostly mere bits, pasted down in a huge folio volume. I suspect this was done by Fraser, and the labels have sometimes been exchanged, so that it requires no little patience. Some of the things I most wished to see are not in the collection, and there are several in the collection which are not mentioned in the "Flora." You would laugh to see what some of the things are that have puzzled us: thus, for instance, his "Cucubalus polypetalus" is Saponaria officinalis! His "Dianthus Carolinianus" is Frasera! in fruit. I will soon send you my notes on the collection, or a copy of them. Bentham looked over the Leguminosæ, Labiatæ, etc., with me. I have had two sittings at Pursh, but have not yet finished; I hope another day will do it, but am not certain. I shall still require about three days more at the British Museum, two at the Linnæan Society, and one at Lindley's. An evening or two at Bentham's will suffice to certify his Labiatæ, Scrophularinæ, etc. I must also have a day with Brown, if I can get it at his own house. I hope very nearly to finish this next week, if life and health are continued....

February 12, 1839.—I am fearful even another day will not see the end of Lambert's collection, and I suspect a week is none too little for the British Museum. Lady Charlotte Bury came into Lambert's and had a long chat with him; such a pair of originals! She is to dine with Lambert on Sunday, but stipulated early, as she always made it a point to read prayers to her servants on Sunday evening!

February 13, Wednesday evening, or rather one o'clock, Thursday.—Rose and breakfasted at eight, which is become my regular practice; started for Lambert's at ten, where I worked incessantly till five P.M.; returned to my room; dressed; went to the City, where I dined, and about eight o'clock arrived at Ward's, whose microscopical party this evening was given chiefly on my account. Some eight or more splendid microscopes were in active use when I arrived; and the greater portion of the chief microscopic people were there. I was introduced to Stokes, Solly, Powel, Bowerbank[70].... Also Mr. Quekett, whom I knew before, and several amateurs, such as Boott, Bennett, Bentham, Don, were present. It was a feast to me, you may be sure, and I acquired some useful knowledge, and saw some strange things: the infusoria in flint; queer fossil woods, which are all the rage here, and are extremely curious; fibrocellular tissue, the most beautiful thing you can imagine. One of the best of the microscopists, Mr. Bowerbank, gave me one or two curious microscopical objects, which he had mounted for himself, and made an appointment with me and another friend to meet him on Monday evening next, to examine his microscopes and curious objects more quietly and at large than could be done in a crowd, and to prepare some specimens for me. Mr. Reade, a gentleman who was invited, but was prevented from attending, was so kind as to send me a copy of his paper on the Infusoria and Scales of Fishes found in Flint, with proof impressions which are far superior to those in the "Annals of Natural History." ...

Tuesday evening, February 19.—Three days have passed since I have written a line for you. This suspension was occasioned by my late hours last night. After spending the morning at the Horticultural Society, then going into the City, where I dined, then going far out on the Mile-End Road to deliver a letter intrusted to me by Mr. Scatcherd, then returning as far as the Bank, I went again, partly by omnibus and partly on my legs, almost as far in the northern outskirts of the town, to spend an evening with Mr. Bowerbank, one of the best microscopists in London, who owns the best microscope. I found so much to see that I did not get away until past twelve, and then I had a walk before me almost the whole length of London,—from New North Road to Charing Cross. I had an opportunity of seeing, what was especially promised me, the camera lucida applied to the microscope; an invaluable invention for an awkward person like me, as I am convinced I could with a very little practice turn out very fair outline sketches of objects I might be examining. I acquired much information on various subjects; saw some most curious and unique specimens of vegetable structure, and particularly of fossil fruits, of which Mr. Bowerbank possesses an invaluable collection; capsules, which we broke open, and examined not only the seed, with its testa, raphe, and funiculus, but even the pulp which surrounded it. I looked at many of his specimens of recent and fossil wood, at his unrivaled cabinet of British fossils, and when our party broke up, there was still so much left that we made an appointment for another evening.... Mr. Bentham, Mr. Brydges, and I went to the Linnæan Society; the president, the Bishop of Norwich, was in the chair,—an amiable old gentleman. Boott, Yarrell, Ward, Royle, Forster, et multis aliis, were present. Mr. Forster[71] invited Dr. Boott and me to fix a day to visit him at his residence, some miles in the country, and dine with him. He is greatly esteemed, and is said to be one of the most kind-hearted and benevolent of men. I am now engaged, I believe, for every day and evening of this week, and half of next, and am busy enough, I assure you....

Friday evening, February 22.—I ought hardly to use the date of Friday evening, as it is close upon one o'clock of Saturday morning. But I must not neglect my journal, and shall therefore give you a few hasty lines ere I prepare for rest. I passed yesterday morning at the British Museum, that is, until near three o'clock. I then hurried to my lodgings, snatched a hasty dinner by the way, and went to the House of Commons, Mr. Bentham having, through Dr. Romily, the speaker's clerk, procured me an order of admittance within the body of the house, where I had the finest opportunity for hearing and seeing. There was nothing very important brought before the house, yet on different subjects nearly all the leading officers of the administration took the floor, Mr. Rice, the Chancellor of the Exchequer, Lord John Russell, who is evidently a man of most ready talent and tact, Lord Palmerston, Lord Morpeth, the new member of the cabinet, etc. I was exceedingly amused by the manner in which Lord John Russell worsted a Colonel Sibthorpe, an opposition member, who moved certain resolutions relative to Lord Durham's expenses, couched in an offensive manner, and made a still more objectionable speech. Lord J. Russell, in very placid manner, set him out in such a ridiculous light, that the gallant colonel first lost his temper completely, and then lost his point, being obliged to withdraw his own resolutions. I heard also, for a moment, Sir Robert Peel, Dr. Lushington, Mr. Hume, and others too tedious to enumerate. As to general decorum, or the manner in which members often treat each other in debate, I don't think we have much to learn....

I spent this morning at the British Museum; dined with Mr. Putnam at a chop-house, and went to spend the evening at Mr. Quekett's. I found, instead of having the evening alone as I expected and wished, that he had invited several friends, most of whom I knew. Still, after tea the microscopes were produced, and I had the opportunity of examining very many curious things.

If they don't get out of my head in the mean time I will try to mention some of them to Dr. Torrey when I go on with my letter to him. As eating is a very important matter here, we had a magnificent supper at half past ten, and it was near twelve when I left, with a walk of four miles before me....

Saturday evening.—This has been a busy and somewhat interesting day with me. I rose early, went down to Bentham's to breakfast, stayed until eleven o'clock, and then went up to Brown's house to spend the morning, according to previous appointment. We talked profound botanical matters, and Brown not only amused and interested me, but gave me much valuable information. He talks of visiting America, possibly next summer, and I have promised to plan him a route. I left him about four o'clock, returned to my lodgings, dressed hastily, took a Kensington omnibus, and reached old Mr. Menzies' little place at five. Mr. Ward, who was to meet us, was not there. We left at half past ten, and walked all the way back, about four miles. So here I am safe again. I read over the doctor's short letter again. I am trying to imagine how Herbert looks now. He has probably changed very much since I parted from him. I have a very especial love for that little fellow.[72] I must find time to write to the girls, yet fear I shall scarcely be able until I have left London. Tell them I think of them daily even if I cannot write them. As to M's French letter, it is not due until I get to France; but that will, I trust, be soon. Adieu. Good-night.

Sunday, February 24. I was fortunate this morning in being able to hear a man I had heard spoken of, and of whom I had formed a high opinion: the Rev. Thomas Dale, Vicar of St. Bride's, who also preaches in the evening at St. Sepulcre's. He preached from the first part of Luke vii. 47: "Her sins, which are many, are forgiven; for she loved much." The discourse was truly evangelical and impressive. He is the best preacher I have heard in England next to Mr. Noel, and is more eloquent and striking in manner than he, but has not the gentle pathos and sweetness of Noel....

Tuesday evening, February 26.... Met Mr. Putnam[73] at half past four. We had arranged beforehand that he should attempt to procure some orders for admittance to the House of Lords, and that we should go down together. I found he had been successful, having sent his clerk with notes to some half dozen peers in order to make sure, and he thus obtained more orders than he wanted. For me I found he had addressed a note in my name to the Bishop of London, who very promptly sent me an order of admittance.

We set out accordingly. The room which is occupied by the House of Lords temporarily, until the New Houses of Parliament are built, is inferior in size and accommodation to that of the Commons; indeed there is nothing about it at all remarkable. There was no business of very absorbing interest before the House this evening, and it adjourned as early as eight. Still I had the good fortune to hear nearly all those speak that I particularly cared for except Wellington (who is sick) and Earl Durham. I heard a long speech from Brougham and a very good one, except that he took occasion to trumpet his own good works. There was some fine sparring between an Irish lord I do not remember, Lord Roden, Lord Westmeath, and Lord Normanby, the late viceroy of Ireland, a young man apparently, and a man of talent, Melbourne, and Minto; the lord chancellor, Denman the chief justice, Sir James Scarlett, old Lord Holland, etc., also spoke. The word "lengthy," which was not long since called an Americanism, seems to be pretty well naturalized, as Brougham used it several times, and Scarlett more than once. Lord Palmerston the other evening used the word " disculpate" instead of "exculpate," which I fancy is rather modern English....

Friday evening, 12 o'clock, March 1.—I have just returned from a most pleasant evening and day, as I may say, spent at Mr. Forster's beautiful residence on the border of Epping Forest, Essex (Woodford), about ten miles from here. He is an old man, a banker, one of the oldest vice-presidents of the Linnæan Society, one of the most kind-hearted men, exceedingly beloved. He lives in an elegant but very unostentatious way, in a most beautiful part of the country, the very perfection of English scenery. He is said to be extremely benevolent, and to do a world of good....

Saturday evening.—Immediately after breakfast this morning I went down to Bentham, whom I had not seen for a week; spent two or three hours there, returned again to my lodgings, went to the City, took an early dinner with Mr. Putnam, and then we went together in an omnibus to Hackney; saw Loddiges' extensive collections of fine plants again, lovely Orchideæ. The Camellias, of which he has a large house filled with magnificent trees, were not yet in bloom.

... We walked across this eastern part of the city down to the Tower, entered the gates and walked over the grounds. It was too late to get entrance to the armory or any of the interesting places, as the light was beginning to fail. I went back to Mr. Ward's, at Well-close Square, according to promise, to name some plants for him, but Dr. Valentine,[74] a most ingenious vegetable anatomist and microscopist, being in town (had previously met him at Lindley's), Mr. Ward had foregone his own advantage and invited Valentine and Quekett to meet me with their microscopes, so that the evening was very instructive to me, which I had not anticipated. Mr. Ward seems to have taken a fancy to me, for I can hardly imagine that he takes so much pains to oblige every one, absorbed as he is also in medical practice. He presented me with a beautiful botanical digger of fine polished steel, with a leathern sheath, which I suspect he has had made on purpose for me; though I don't know why he should have thought of it. Mrs. Ward was inquiring about the Abbotts and their works, one of which she had, which makes her wish for more. I am often asked about Mr. Abbott, whose works seem much more generally known here than those of any other American religious author. I must find some for Mrs. Ward.

Sunday evening, March 3.—I went this morning to hear, perhaps for the last time, Baptist Noel. The sermon was from the last three verses of the same psalm (Ps. ciii.) from which he has preached on the former occasions when I have heard him in his own church; and truly a good sermon it was. I have told you that the chapel is a large one. Yet it is so well filled that I have always had some difficulty in getting a seat, and to-day I actually stood near the pulpit during the whole service and sermon. But it is worth while submitting to some inconvenience. In the afternoon I walked up to Tottenham Court Road, and looked up the chapel built by Whitfield, the scene of his useful labors in London. If you read, as I think you did, Philip's "Life of Whitfield," you must take some interest in this place.[75] I found the chapel a large but outlandish building, with an inscription over one of the entrances, stating that the building was erected by George Whitfield. Within is a tablet to the memory of Mrs. Whitfield, who is buried here, and a monumental inscription to Whitfield himself (which I regret I did not copy), mentioning the date of his death at Newburyport, near Boston. The preacher this afternoon (for I believe there is more than one who officiates here) was the Rev. Mr. Wight, who gave an impressive, practical sermon from the concluding clause of the last verse of Romans viii.: "The love of God which is in Christ Jesus our Lord." It was, I think, rather above his audience, which I am sorry to say was exceedingly small. Indeed I hope it is generally better filled, but I should not have expected so great a falling off in the attendance of plain unfashionable people in the afternoon. These Whitfieldians are, one would think, farther separated from the Established Church than Wesleyans (which was certainly not the case in Whitfield's time, who refused to take any steps to establish a sect apart from the Church of England); for in the Wesleyan chapel I attended the liturgy was read, but here we had none of it. Only last summer I read a biography of Whitfield with much attention; and it was very interesting to worship in this chapel of his. It recalls more interesting associations than Westminster Abbey or any vast and splendid cathedral. But I must bid you good-night, purposing to rise early and have an hour or so before the pressing business of the day is commenced to write another sheet to you and our good Dr. Torrey, to whom I have so much to say, if I could ever find time for it.

Friday.—I have been to-day at the British Museum, studying from the specimens of Plukenet, Catesby, Miller, etc., etc., the authority for old Linnæan species in Ilex, Prinos, Eupatorium, etc. It is slow and tedious work, and I shall not have time to do so much of it as I could wish. Brown told me to-day about Petalanthera. It is Cevallis, Lagasca, Hortus Matritensis, and very probably his species, even C. sinuata. It came from New Spain. You will see Lindley is all astray about the genus, and no one knows its affinities even, but Brown. Lagasca himself refers it to Boragineæ. It is true Loaseæ. I was this evening at Bentham's, and found he had a specimen of C. sinuata from Hooker, collected by Brydges in Mexico, I think. I have asked Brown to give us some notes on the subject, a generic character, etc., that we may publish a little from his own pen. I am to spend a day with him next week, and I will try to get something out of him. He hinted to me some days ago that he knew something about Cyrilla, but I could not get it out of him. I'll try again. He tells me he has a character to distinguish true Rhexia, which has escaped Don, De Candolle, etc. We must find it out. Bentham has given me his "Scrophulariæ Indicæ," and the three last parts of his "Labiatæ;" I have bought the rest (£1 2s. 6d.), and last evening we looked over his North American specimens, and the notes in his copy. He gave me also, the other day, the only published part of the "Plantæ Hugelianæ" and a few other pamphlets. He is a liberal soul.

I have got so far behind in my botanical news that I despair of bringing up arrears, and must leave very much to tell you in propria persona, if we meet again. I fancy I have not very much new to learn on the Continent about microscopes and modes of working. I have seen much of all the best people here, last not least Valentine, who lives in the country, from whom I have derived much useful knowledge. He works to some account, which can't be said of most here, who, though they have the best instruments in the world, don't turn them to any important account. As to Sullivant, tell him to have great patience. I can get him a capital simple microscope by Ross for six guineas, but I want to get as useful a one for him cheaper, so I shall wait till I have been on the Continent, I think. My plan is to purchase at Paris for him, where the low powers are good as can be, and supply a lens or two here....

Chapmannia (!) exists in Bartram's old collection here, which you saw at British Museum, and some other very lately published things.

I bought a copy of "Flora" for Bennett the other day, thinking it worth while to offer him something, as I was taking up much of his time. To-day he gave me a copy of the published part of the "Plantæ Javanicæ Rariores," (£2 10s., plain, is the publishing price), an invaluable work, containing very many notes and observations on various genera, etc., both by Brown and himself, which it is quite necessary we should see. The notes

I have made for the last few days are not now before me, so that I cannot now give you any remarks. There is no one thing of very considerable importance, but much small matter. By the way, let me say that Bennett thinks that Brown thinks Romanzovia to be hydrophyllaceous! Bentham would give something to know this, but I shall keep it to myself. I have made out the remainder of Pursh's doubtful Arenarias and Stellarias front the Banks herbarium. The parcel of Solidagos, etc., sent to care of Mr. Putnam, I am glad to say, came to hand. It did not arrive until last week, however....

Monday evening, twelve o'clock.... As I sit down to tell you what I have been about to-day, my thoughts cross the wide wave that separates us, and brings me back to 30 MacDougal Street, and to the time when, returning from town, I used to present myself before you, give an account of my proceedings, tell you perhaps some news about that ill-fated expedition of which you were so sick of hearing; how it would certainly sail in a month, or something just as likely. When thinking of this long separation, I console myself with the idea that it is better than if I had gone there. In that case I should now have been your antipodes. Now there are only some four or five hours of shadow between us. And, sluggard as you call me at home, I am up in the morning two or three hours before you. Tell that to the girls for a wonder! I left my room this morning at eleven, walked to Portland Place, called on the American minister, who being unwell I was furnished by the secretary of legation with what I desired, namely, a passport. This I left, as the manner is, at the office of the French embassy, that his majesty Louis Philippe may have fitting notice of the honor that is to be done him, for the king of the French is, it seems, rather particular abut such matters, and it is a pity not to oblige him, especially as you can't help yourself. This being done I went on to the Linnæan Society, and by working at the full stretch of my powers contrived to get through the Linnæan herbarium (skipping a few genera now and then) about six o'clock. Returned home pretty well fatigued, took some tea and toast, called upon Bentham, whom I found writing letters of introduction for me. I have them now before me. They are addressed to Seringe at Lyons; Requien, Avignon; Lady Bentham (B.'s mother) at Montpellier, with request to make me acquainted with Dunal and Delile; Moretti at Pavia; Visiani at Padua; Tomasini at Triest; Unger at Gratz; Endlicher at Vienna; Martius and Schultes at Munich; Reichenbach at Dresden; Pöppig at Leipsic. These, with what I have already from Hooker, Arnott, Greville, Boott, etc., with a few that I expect at Paris, leave me little to wish for in this respect. About ten o'clock went to Mrs. Stevenson's party. It was not a very large one, and in no way especially remarkable. I found there of course the Bootts (three sizes, viz., Mrs. Boott the grandmother, Mrs. Boott the mother, and Miss Boott the daughter) and so of course I was upon good footing. Our minister lives in neat but by no means splendid style, quite enough so for a republican; and Mrs. S. is very lady-like and prepossessing in appearance. Mr. Stevenson did not make his appearance. Of course, I did not stay long.

TO JOHN TORREY.

Poor Hunneman died yesterday, after a short illness. I have spent much time evenings with Mr. Valentine, whom I like extremely. Excepting only Brown, he is the best microscopical observer in Great Britain. He cares little, however, for proper systematic botany, for which I am sorry. He has shown me some curious things.

I have learned from Brown the character he observed in our species of Rhexia, that is, the true genus Rhexia: the unilocularity of the anthers....

Tuesday evening, March 12.—After a hard day's work I finished on Monday evening with the Linnæan herbarium, which I found more interesting than I expected and more satisfactory, as it is in really good state, carefully taken care of, etc. I had some very good notes to make. I assure you I feel much gratified to have studied this collection, which, with the Gronovian, enables us to start fair as to Linnæan species. Do you know that Acer saccharinum, Linn., is A. eriocarpum (spec. Kalm)! Look at Linnæus "Species Plantarum" (which you have not, unfortunately, though it is the most necessary of books; you will receive it at the same time as this letter or nearly) and you will find that the description is all drawn from Eriocarpum.

I took what time I could to-day for the Gronovian plants and a few of Plukenet's, etc., but was unable to finish; will go to-morrow, for I shall work to the last moment.

I have been tempted to buy a collection of Hartweg's[76] very fine Mexican plants, which being collected far in the interior of north Mexico are very North American, and quite necessary, I think, for us. They will reach you with the other parcels. Be careful about the little labels with the numbers stuck on. Bentham will publish them presently....

Professor Royle, as the agent of India people, I believe, offers me seeds from Himalaya Mountains, received, and still to be received, from the government collectors, in exchange for those of useful and interesting North American plants, which they are desirous of introducing into India. But as I can't attend to it until another season, he kindly offers to send to you a portion of the seeds just received, and to ask you to distribute them in such way as will be most useful, and ask those you give them to (say Downing, Hogg, Dr. Wray, Dr. Boykin, etc., and some one in the valley of the Mississippi or Arkansas) to collect seeds of trees, etc. (you can suggest what would be most desirable), and send them to London, whence they will be sent in the mails overland to India. As I fear I shall not see Royle again I shall write him a note, telling him, as I promised, how to send to you.

I saw Dr. Sims' herbarium, at King's College. I want to look at it to certify a few early "Botanical Magazine" plants.

Brown came to the museum this morning with a copy of a curious late paper of Schleiden (which I had seen before) on the Development of the Embryo, with a parcel of his own notes on the same subject made in 1810, 1812, 1815, etc., which did not altogether correspond. Brown thinks much of Schleiden as an observer. He read me many of his old notes, and the subject took him to speak of his discoveries with regard to the embryos of Pinus. To explain to me as he went on he drew the diagram on the inclosed slip of paper, and pointed out to me how to observe in our species of Pinus. This will refresh my memory as to all he told me, so pray keep it safely. There is much very curious matter now afloat about the process of impregnation and the early development of embryo, which I am accumulating, as much as I can, for future use. Pray tell Dr. Perrine that the gardeners and botanists here insist by acclamation almost that there is no such thing as acclimation in the vegetable kingdom.

What a pickle the Linnæan Ascyrum is in! I wish I had room to tell you.

TO MRS. TORREY.

Tuesday morning, two o'clock A.M., March 14, 1839.

I have just finished packing up, being about to start for Boulogne in steamboat at nine o'clock this morning, and I must now hastily close my letters. This, or rather yesterday, has been a busy day with me. I started in the morning to have a look at a few more things of Pursh's at Lambert's, but he kept me longer than I liked. He found somewhere a small parcel of plants collected by Eschscholz in Kotzebne's voyage, who sent them to Lambert. Lambert gave me all the North American ones, few to be sure, but interesting. From Lambert's I returned by way of the Horticultural Society, to bid good-by to Lindley and Bentham, but the latter insists upon coming up in the morning to my lodgings to see me off. I have made a fortunate acquisition for him. He told me he saw, a few days ago, at an auction some copies of Richard's fine work on the Coniferæ, but an engagement at the time prevented him from staying to buy a copy of the work for himself, which he imagined would be sold cheap. Mr. Putnam found out who bought up these copies, and obtained one at nearly the price at which they were sold. I shall have the pleasure of presenting it to Bentham this morning when he calls. I went to the British Museum, worked hard until four o'clock; but was not able quite to finish, so I left my copy of Gronovius, in which I was making notes, with Mr. Bennett to keep for me until my return in the autumn, and took leave of Brown and Bennett. Went to Dr. Boott's; saw Mrs. and Miss Boott, who insisted upon giving me a note of introduction to a friend of theirs in Florence; went to the City, dined with Putnam, down to Well-close Square, took my tea, and bid good-by to Ward and family, and Mr. Quekett....

TO THE MISSES TORREY.

PARIS, March 18, 1839, Monday evening.

I am now at the Hotel de l'Empereur Joseph II., Rue Tournon, près du Palais du Luxembourg. Here I have been established for about half an hour, and my first business shall be to fill this sheet for you. I suppose I must begin at the beginning and tell you how I came here. Voilà. I left London at nine o'clock in the morning of the 14th inst. (Thursday), stopping on my way to the steamboat which was to take me to Boulogne, to leave a parcel of letters at Mr. Putnam's office, to be forwarded to dear friends at home. It was a nasty, rainy morning; and our boat was, as indeed I expected, not very comfortable. The cabin was well enough, but much too small for the accommodation of some fifty or sixty persons, and there was no covering to the deck, nor any deck-cabin, except two dirty little places for the poorer passengers, who were not allowed the use of ours; so we had our choice the whole day between the soaking in the rain upon the deck and the close atmosphere of the crowded cabin. Of course I was vibrating between the two dilemmas the whole day, but took as much pains as I could to keep dry. The only thing I saw worthy of notice as we went down the Thames was Greenwich Hospital, of which I will perhaps send a print. I should add also chalk cliffs, for I never before saw rocks and hills of chalk. In the afternoon, as we had fairly got into the Channel, a thick fog came on. The captain lost his way and seemed in fear that he should run the boat upon the Sands, so he dropped anchor about five in the afternoon. We were to have arrived at Boulogne at nine that evening. But as I saw there was no great chance of our moving for some time, I set about making amends for my loss of sleep the previous night. I took possession of two thirds of a hard sofa, and, wrapped in my cloak, was soon in a comfortable doze. I awoke late in the evening; and such a sight as there was before me! It seems that there were no accommodations for sleeping on board, or next to none, and the passengers, men, women, and children, were indiscriminately but thickly strewn over the sofas, chairs, and even over the whole floor, with portmanteaus, great-coats, and whatever they could find for pillows, attempting to secure such rest as they could,—some sixty persons or more crowded into a space not larger than the cabin of one of our ferry-boats....

But I was too drowsy to mind it much, and soon fell asleep again, but awoke in the morning with swollen eyes and complaining bones. The boat was moving again, and it was raining as hard as ever. The distant coast of France soon came in view, and at half past ten we were landed at Boulogne. We were escorted to the custom-house; what baggage we had brought in our hands was closely examined by the officers, an ill-looking, vagabond set; our passports were taken from us and provisional ones given, which permitted us to go on to Paris, and for which we each had to pay two francs; we were then allowed to go to a hotel and get our breakfast, a privilege which most of us were not slow to avail ourselves of. I made a hearty meal of cold roast beef, café au lait, excellent bread, and delicious butter. The two last I have found ever since I have been in France. I gave my keys to the commissionaire of the hotel to get my luggage through the custom-house, and, my place being taken in the diligence for Paris at two o'clock, having nothing else to do, I went to the custom-house to see the examination of the luggage. Lazy custom-house officers and gendarmes were lounging about, while heavy carts loaded with baggage were drawn up from the boat by women!—and this while it was raining hard, and the poor creatures were without hats or bonnets, and had only a handkerchief or a bit of cloth tied over their heads. So much for this self-styled most refined and polite nation! I noticed the poor things when their task was done and they were waiting to convey the trunks, etc., from the custom-house to the various hotels. Some were chatting in groups, apparently quite content with their lot; a few were sleeping, and many, with the characteristic industry of their sex, produced their knitting-work from their pockets and were busily employed at a more appropriate and feminine employment. I was amused at the strictness with which three exceedingly unpleasant-looking fellows searched all our baggage, that of the ladies not less than that of the men. Little parcels were opened, dirty linen was overhauled and most minutely inspected; the whole scene would have made a fit subject for the pencil of Hogarth. My traveling-bag was examined from top to bottom, and I began to fear that my trunk, which I had packed with care, would be sadly deranged, but they contented themselves with cutting open a packet of seeds I was taking from the Horticultural Society to De Candolle, and with seizing as a great prize my rather formidable parcel of letters of introduction. This was near causing me to be detained until the next diligence; but the commissionaire succeeded in getting them sent up to the inspector in another part of the town, upon whom we called, when after due explanation had been made, and one or two of the letters read, they were formally delivered back to me.

I can tell you what a French diligence is like. It is just like one of the railroad cars (about three apartments) of the Harlem railroad, for example, mounted on coach wheels; the horses are small, lean, shaggy, and ugly;

some seven of these beasts are fastened, three abreast and one for a leader, with ropes to the said diligence; but how such beasts contrive to draw such a cumbrous vehicle, loaded with seventeen persons and their baggage, besides a driver and conductor, I don't well understand, although the beasts are changed every five or six miles; but somehow we got over the ground pretty fast, and came to Paris, over one hundred and forty miles, in a little less than thirty hours, although it rained all the first day and part of the second, and the roads were extremely muddy.

We arrived just before nightfall at Montreuil, a fine old fortified French town situated on the summit of a hill and overlooking a broad valley, which in summer must be quite beautiful; here we dined, and were charged four francs each for dinner, besides sous to the garçon. I slept pretty well in the night, during which we passed Abbéville, where there is said to be a fine church. We breakfasted at the queer old town of Beauvais, where there is a fine cathedral, of which I had a pretty good view. My breakfast (déjeuner à la fourchette, which is the next thing to a dinner) cost three and a half francs, for on this route you meet with very English charges. I wished to say something about the country, but have not room. Suffice it to say that we passed through the town of St. Denis late in the afternoon, where I did not even get a glimpse of the very ancient cathedral, and arrived at Paris just before nightfall. After dinner, in company with a fellow-passenger, a young Englishman, I gratified a long-felt curiosity by strolling through the Palais Royal and some of the principal streets of Paris. On Sunday I attended church in the morning (after a vain attempt to find the American Chapel) at the Rev. Mr. Sayer's English Episcopal Chapel, where I heard a good sermon; and in the evening at the Methodist Chapel, where the Rev. Mr. Toase preached a truly excellent discourse from Jeremiah viii. 13. All the shops were open just as on any other day, and the gardens and parks were all crowded. This morning I went down to the Jardin des Plantes, stopping by the way to see the ancient church of Notre-Dame, where I heard a portion of the Catholic service chanted.... At last, after looking at many other buildings and objects of curiosity, about which I will tell you more presently, I reached the garden, found Decaisne, who could speak no English, and I almost no French; so he took me to Adrien de Jussieu, who makes out to speak very tolerable English, and to understand me pretty well. I left soon to call on Mr. Webb,[77] who is an Englishman, for whom I had a letter from Hooker; thence after looking in vain for "appartements garnis " in Rue de l'Odéon, Place de l'Odéon, etc., I secured my lodgings here, where I shall be obliged to hear nothing but French, and where I hope I may catch some of the language, and after dining at the ordinary at the Hôtel de Lille, where English is spoken, I transferred myself to my present quarters. But my sheet is full. I will give you another very soon. Till then, mes chères petites sœurs, adieu.

Wednesday evening, March 20.—I must continue my letter to you on a large sheet of thin French paper, else I shall have a larger bill of postage to pay than will be altogether convenient when I send to Havre. I did not write last evening; I had no fire in my room, and after running about all day over streets paved with little square blocks of stone, which it is very fatiguing to walk over, I came home fairly tired, and went to bed soon after nine o'clock. Except calling on M. Delessert, for whom I had a letter and a small parcel from Hooker, and whom I did not find at home, I spent the whole day in looking about the town, seeing sights, etc. My first call was at the Louvre, a large and splendid palace, where I spent an hour or two in the vast gallery of paintings, which fill a very large salon and a long gallery, I suppose five hundred or six hundred feet long, connecting the Louvre with the palace of the Tuileries....

To-day I have been wholly occupied at the Jardin des Plantes. Fortunately for me Jussieu speaks a little English, so I can get on with him pretty well. But you would have been amused at the attempts which M. Decaisne and M. Gaudichaud[78] and myself made to understand each other. Still more amused would you have been to see how I managed to make a bargain with a bookseller for a few books I wished to purchase. I feel the want of French sadly, and have no time for study.

Thursday evening.—I have been again occupied the whole day at the Jardin des Plantes, and went at six o'clock to dine with Mr. Webb to meet M. Gay.[79] Webb had taken care to ask an English student also, who speaks French much better than he does English, who sat between Gay and myself and interpreted when it became necessary. But Gay speaks a little of what will pass for English, mixed here and there with French, so that I got on very well indeed.

Gaudichaud was also there, a very interesting man if one could talk with him. We were kept rather late, so that it is now past twelve, so I must bid you good-night.

Monday evening.... At three o'clock I went to the Institute. I found that the room was already crowded. I inquired for Jussieu and Brongniart, the only members I could think of that I knew, but they were not there and therefore I could not get in. After some time Jussieu came in. But it was then too late, so I lost the object for which I had given up half the day. Jussieu, however, took me into the library, which is worth seeing. I employed the remaining hour or so in purchasing some prints of remarkable buildings, etc., in Paris, and I was also tempted to buy a few engravings from some of the great masters. After dinner I went to Mr. Webb's, where I looked at plants for a few hours. He gave me also some autographs of celebrated botanists, and a few old botanical books....

Friday evening, March 29.... The Garden of Plants was nearly on my way home; so I stopped there, worked for an hour (till five o'clock), went home (home, indeed!), took my dinner, found myself most thoroughly tired as well as hungry, having had no breakfast but a small roll of bread I obtained near the cemetery; had a fire kindled in my room, and commenced writing to you. Just now the little daughter of the concierge, a little girl of six or seven, who often waits upon me, has brought me a cup of coffee, which I have enjoyed greatly, and now feel much restored. French children are all pretty and graceful, and I am making the little girl's acquaintance as fast as I can; for it is difficult for me to understand her (it seems odd to hear such a little thing speak French), and in answer to some of my attempts to speak French to her, she answers, "Je n'entends pas anglais, monsieur."

What great lies the French newspapers tell! Yesterday morning the paper I was reading at my breakfast stated that one of the gardeners who had charge of the bears at the Jardin des Plantes descended into the inclosure for some purpose, and was seized by the bears, killed immediately, and almost eaten up before help was obtained. So when I arrived at the garden I of course spoke to Decaisne about it, who was greatly surprised, for it seems the story was entirely a fabrication.

I see I have at length filled this large sheet, so I must say adieu for the present, but hope to-morrow evening to begin another. Ever I remain,

Your attached,
A. G....

TO MRS. TORREY.

Wednesday evening.... There is little danger of my being spoiled in Paris by being overpolished. In London one must take care to be always comme il faut. There I took pains to keep myself rather spruce, which I have continued here from the mere force of habit!!! But gentlemen in Paris dress anyhow; they don't pay half the attention to the matter it receives in England; with the ladies it is perhaps different, but here I scarcely ever see ladies except in the streets or shops and restaurants! At the houses of botanists I have only seen Mme. Gay, a very plain and good-natured Swiss lady. As to parlez-vous-ing, it is not such an easy matter, I assure you. You would laugh most heartily to see me in the botanic gallery of the Jardin des Plantes, endeavoring to carry on a conversation with Gaudichaud or Decaisne; the former of whom can scarcely read English, and the latter can speak only a dozen words. I get out, with no little difficulty, a few sentences of such French as has not been heard since the days of King Pepin, I am sure; and when that fails me I write in English, which Decaisne can read, and make him write in French in return, or else for short sentences speak very slowly and distinctly. From my ignorance of the language I am obliged to take great pains when I wish to purchase anything from the shops; for it is customary to put on an additional price to English customers. Fortunately my complexion and the style of my countenance are so far French that before I speak I am generally taken for a native, and I sometimes manage to make purchases without saying a word beyond a monosyllable. So I have to be very careful to avoid being cheated; but I am every day acquiring more knowledge and experience.

I have been seized with a mania for collecting prints on a small scale, and shall send home some very good ones,—to adorn my parlor and study at Michigan, of course! There are astonishing quantities to be found here. I am endeavoring to get all the portraits of botanists I can, and from this I have been led to pick up ancient ones, which show the early state of the art or old-fashioned costumes, etc., and also a few choice engravings from the old masters; but most of these I can obtain better in Italy or Germany. Tell Dr. Torrey not to be alarmed, for I shall not spend much money upon them.

As a general thing Paris is not very beautiful. But there are some magnificent sights, I assure you. At odds and ends of time I have already seen most of the ordinary sights which attract the attention of travelers, but must leave all account of them for the journal from Paris, which so far is addressed to the girls, though I fear it will scarcely interest them or any one else....

Decaisne has given me separate copies of his papers. He is now publishing a most splendid (botanically speaking) memoir upon the order Lardizabaleæ, in which I see he has found out some things which have been known to Brown only, for a long time. He will give us copies, I dare say. He is one of the best botanists here. I like Gaudichaud also very much....

I have just finished the examination of Michaux's herbarium, which has proved worth looking over. I shall write the doctor more particularly, indeed have already begun a letter for him. Mr. Webb showed me last evening a letter from Hooker, which contains a good deal of botanical intelligence for himself and me. The British Antarctic expedition, he says, is to sail positively in August, and Joseph is to go. I wonder if they will be two years or so in getting off!...

TO THE MISSES TORREY.

PARIS, April 1, 1889, Monday evening.

MY DEAR GIRLS,—It is rather late, and I have no fire in my room, to which I have just now returned, but it is nearly comfortable without one, and so we will have a few words together before I sleep. My last and long sheet was closed, I think, on Friday evening. On Saturday my morning was spent as usual at the Jardin des Plantes; returning from whence I looked along the shops and so on to the Pont du Louvre, which I crossed; passed through the Palais Royal at the most busy season, when it is all lighted up splendidly, and dined at the Restaurant Colbert at half past seven. I am patiently exploring (I should say eating) my way through the mazes of French cookery, and am trying to select from the complicated bill of fare the more peculiar and national dishes, some of which are excellent, others so-so, or very poor....

To-day I have been again at the Garden, working as hard as possible, since I have so little time remaining. I dined at half past six at one of the famous restaurants, just to see how it was managed, and returning spent the early part of the evening with Mr. Webb, who lives near me.

On my way from the Garden, I stopped at another church. I believe the only remaining one of large size and much interest which I had not already seen.... It is called St. Severin, and is very old, having been built in the year 1210.

This is the first of April, and a fine spring day it has been, though the season is little more advanced than at New York. In two weeks I must be again upon the wing, and shall soon meet the summer. I want to see the south of France and sunny Italy. Adieu.

Tuesday evening, April 2.—I intended to have had time this evening to write several letters, but Decaisne has been with me, and did not leave until almost twelve, we had so much to talk about. I have been all the morning at the Garden; have worked very hard, indeed, and have nearly finished there. To-morrow is like to be a broken day, as I have made an engagement to see Dr. Montagne[80] and his microscope at twelve o'clock, which will take an hour or two out of the very best part of the day. I will try to turn the fragments of the day to some account. But now good-night.

"To each, to all, a fair good-night,
And pleasing dreams, and slumbers light."

Monday evening, April 8.... Saturday was a little more diversified. I went at eight o'clock in the morning to Professor Richard's,[81] who lives near me, examined some plants of Michaux, then took my breakfast, went to the Garden for three or four hours, but returned at two o'clock to see the Chamber of Peers in session, M. Gay having provided me with a ticket of admittance, which procured me a very good seat. The members all wear a kind of court dress, the military peers swords, and those who have them display the insignia of the order of the Legion of Honor, and so forth. Several new peers were admitted, but before they were introduced, a number of peers made some remarks which could not have been very flattering to them, the creation of a new batch just at this time having given much dissatisfaction to the old ones. Among others, I heard a little speech from the famous Marshal Soult. Lord Brougham, who is now in Paris, was present. I recognized him across the room by his homely face, which he is in the habit of twitching and contorting incessantly, as if it pained him. He seemed to listen with much attention.

In the evening I paid a visit to Mr. Spach,[82] looked over plants and so forth until ten o'clock, returned shivering with cold, for the weather here is like March in New York. I am now sitting by a large fire, and yet I am shivering.

Tuesday evening, April 9.—In the morning went to hear Mirbel[83] lecture at the Sorbonne; he speaks so distinctly that I understood him tolerably well in general. The lecture-room is old and incommodious, rather better, to be sure, than the accommodation for the students of the university in the olden time, when they used to sit upon straw spread in the streets, but certainly not very fine. I went afterward to the Ecole de Médecine; heard the professor of anatomy for a few minutes; came away, saw two or three books that I wanted in a stall belonging to a shop, priced them; found the price much higher than I intended to give, so I named the price I would give; was amused with the perseverance of the very genteel madame, who reduced her price down to within seven francs of my offer, and then labored hard to make me take them. I advanced one franc, but utterly refused to give a sou more. "Vous n'êtes pas raisonnable," says madame. "Je suis très raisonnable," I replied, "mais votre prix n'est pas raisonnable." So I left the shop, madame very coolly replacing the books on the shelf, with one eye turned toward me to see if I would relent. I had got some distance down the street when the boy came running after me, to say that I might have the books, "mais ils sont très bon marché." So much for the way you are obliged to make bargains here. Went to the Garden, returned to dine here, paid a little visit to Mr. Webb, and must write the remainder of the evening.

Thursday evening, April 11.—My approaching departure makes it a very busy time for me. Let me recollect what I did yesterday. I went first to Baron Delessert's; studied in his magnificent library until about one o'clock; then visited my banker, who is near, drew some money; then to a bookseller to arrange some matters about our "Flora" (which I failed to do); went to the Bibliothèque du Roi, where they have miles of books and acres of manuscripts, but as it was not a public day, I did not see half that I wished. I have made arrangements, however, for a future day. I went next to the post office, and took a place in the malle-post (which is very much quicker than the diligence) for Lyons, to go on Monday; so that the time of my departure is pretty well fixed. I next went to learn the time of the departure of the carriages for Sèvres and Versailles, which places I intend to visit to-morrow. Then I met Chevalier, the optician, by appointment, to consult about microscopes for an hour or two.... Called on M. Gay, with whom I found M. Boissier, a Swiss botanist whom I had often seen at the Garden, and also August St. Hilaire,[84] who returned but a few days since from Montpellier.

On reaching my room at half past ten, I found a note from Mr. Webb, saying that M. Spach had a message for me from Mirbel, and asking me to call if I had time; went immediately, but was too late; Webb had gone to bed. Returned, arranged accounts, etc., and went to bed myself.

To-day I have been, if possible, still more busy; at least I have accomplished more, though I made a bad beginning. The concierge promised to call me at eight, but I awoke myself at nine. Consequently it was past ten before I made my first call, which was upon Mr. Webb, to know when I was to see Mirbel. I called next upon Dr. Montagne to get a letter to the chief curator of the Bibliothèque du Roi, which should afford me the opportunity of seeing this, the largest library in the world, on a private day, namely, Monday, the only public day while I stay being Friday, when I have something else to do. *Eh bien.* I went next to the Louvre, and saw the other and best half of that most magnificent gallery, my passport giving me a ready admittance.... Suffice it to say I saw very much to admire—some things that I greatly admired—very much I did not allow myself time enough to become interested in, as well as many works of the old fellows that one likes to say he has seen.... Again in a cabriolet to the Ecole de Médecine; looked through the museum, which was to-day open to the public; saw for a moment the examination of a batch of candidates for a vacant professorship by concours; also the examination of students in the same way; then I visited the Musée Dupuytren,—a surgical museum of great extent; then went to the Ile St. Louis (opposite the Garden) to call on M. de St. Hilaire; not at home, so I saved a little time. Next to the Garden; looked on my way at the animals, the hyenas, lions, giraffe, monkeys, etc., besides a few large snakes; then called at Mirbel's rooms, who took a great deal of trouble to show me most curious things in vegetable anatomy, but of this I will write to your good papa, who will care much more for it than you. After this I saw Decaisne for a few minutes at the botanical gallery; took one of the young lads with me; saw the mineralogical cabinet and that of fossils, which occupy a new and most beautifully arranged gallery. Here I saw many of the famous things I have heard so much of. In the vestibule to this gallery they are preparing a pedestal for a fine and large statue of Cuvier. I went next to Jussieu's house, talked with him for a few minutes, and bid him good-by. On my way home stopped at Ballière's, the bookseller, to transact some business; home; dined at half past seven; went to Webb's, where I like to go of an evening, as I get a good cup of tea (no common thing in Paris), which, after such a day's work, was very grateful, I assure you; remained until half past nine; returned here, took up my pen, and voici the result; and if I do not write plainly and neatly, it is no great wonder, and I trust you will excuse it, for I have other writing to do also this evening. Besides, I must rise at seven, as I expect another very busy day. On my return this evening, I found a polite note from Delessert[85] accompanying a magnificent present, no less than a copy of three volumes of the "Icones Selectæ." An invitation for Saturday evening from M. and Mme. Delessert came with it. I am already engaged to dinner, at half past six, for the same day.

JOURNAL.

Saturday morning, half past seven.—[After an account of a visit to Versailles, he goes on:] Now bidding adieu to all this most interesting ground, I took up my march, on foot and alone, for St. Germain, distant about four miles. From the heights of Louveciennes I obtained the first view of the Seine and the lovely and broad valley through which it winds. Here I passed the remains of an elevated and striking aqueduct which conveyed water to a royal château which formerly stood in the neighborhood, and also, I believe, to the village of Marly, through which I passed a little farther on. Then descending rapidly, I reached again the banks of the Seine, the terrace of St. Germain being directly before me. It was now three o'clock. The steep hill was to be ascended by a winding road, and being somewhat leg-weary, I stopped a passing countryman's cart; the lad who was driving readily gave me a seat by his side, and thus I rode into St. Germain. The lad was quite intelligent, and answered all my questions (when he understood me) very readily. He set me down close by the château. I gave him ten sous for his trouble, and we parted on good terms with each other. The château of St. Germain, which was a chief royal residence before Versailles was built, is more interesting to us as the place where the Stuarts kept their petty court so many years. It is now converted into a military penitentiary, and I was not anxious to examine the interior, as I am informed scarce any of the original apartments or furniture remain. The exterior is striking, quite of the old style, built of the same red bricks as the central portion of Versailles. What is most worth seeing here is the terrace, a beautiful park, extending for almost two miles along the brow of the high ridge, with the most beautiful view from it of the valley beneath and before you, the hills that bound your view, and the numerous villages scattered here and there. A finer situation cannot be imagined. The Seine, after passing Paris, makes a bold, double turn. The view extends quite to Paris (fifteen miles) though the city is nearly concealed from view, yet you see the grand Arc de l'Etoile distinctly. In the summer it must be surpassingly beautiful. At four o'clock I descended the steep declivity to the commencement of the railroad, took a little refreshment; at twenty minutes past four we started in cars propelled by steam, and in an hour I was in Paris and taking my dinner at the Restaurant Colbert. A pretty good day's work!

Saturday, went to dine at Mr. Webb's; a little party,—a bachelors' party, for Webb is single,—consisting of Dr. Montagne, M. Berthelot, M. and Mme. Ramon de la Sagra, M. Spach and his wife, and a young Spaniard whose name I do not recollect. Webb is quite a polyglot; he speaks French, Spanish, Portuguese, Italian, Modern Greek, and I know not what besides his mother tongue. At half past nine I left, took a cabriolet for Delessert's, where I had been invited to an evening party; found there several botanists and persons I knew. Delessert received me cordially, introduced me to Madame D., who I was rejoiced to find spoke English very well. The suite of rooms thrown open was very splendid, and communicating with the last was a pretty greenhouse, filled with vigorous plants, all in fine bloom; the whole, carpeted and lighted, presented a most inviting appearance. The brothers Delessert are said to be very rich, and I suppose can well afford such an expensive establishment. The party broke up at eleven. Besides tea, which is quite English, though the French are getting more into the custom of using it we had ices, etc., but nothing else. The whole affair was conducted without any parade and in quiet good taste....

Notabilia varia.—Ellimia, Nutt., was described a little before us by two authors under two different names: First by Cambessides in Jacquemont's Travels, under the name of Oligomeris; second by Webb and Berthollet, "Histoire Naturelle des Iles Canaries," under the name of Resedella; Webb has Jacquemont's plant from the Himalaya and his own growing together; they are absolutely the same. I am to examine them soon, but have scarce a doubt they are even the same species as ours. Webb has promised me a specimen. It is also the Reseda glauca of Delile ex Egypto. It is curious that the plant should at the same time be described from almost every part of the world, and not less so that the three names hit upon should have all meant the same thing, namely, a reduced reseda.

I have just spent the evening with Gay. He is publishing Carices in "Annales des Sciences Naturelles:" has hit upon some of Boott's notions; but not all. He is a laboriously minute observer, and will do pretty well, but like Boott inclines to make too many species. He insists upon describing the small form of C. Hitchcockiana from Dr. Sartwell and Kentucky as a distinct species, in which he may be right. He wished to name it after me, but I declined the honor, and have transferred it to Dr. Sartwell, the discoverer, whose name it is to bear....

Delessert received me very kindly when I called on him. I must call again soon, and consult especially his rich library. He showed me a list he had just ordered from New York; among which of course was our "Flora." I should have offered him a copy, but now it is scarcely worth while.... I shall not see De Candolle here. Delessert does not expect him until May. I shall leave the books and parcels for him with Delessert, and make De Candolle take back to Geneva with him all my parcels that I do not wish to take with me to the south.

April 2, evening, or rather April 3, as it is past midnight.—I have worked to-day as hard as I could jump from ten to half past five o'clock at the herbarium général of the Museum de Paris, and have finished. Apart from Michaux's plants, of which they have nearly a set distributed, they are wretchedly poor in North American species; almost none of Lamarck and Poiret. I except the plants given by LeConte, Torrey, etc., which are arranged but not incorporated. The present Gallery of Botany is exceedingly fine and spacious, and well planned. I have gone carefully through all Michaux's herbarium (from your limited time you have made some bad slips in the Carices of Michaux, which Gay, I am sorry to say, has found out), noting all dubious matters to be settled by examination of Richard's set. I have gone through De la Pylaie's herbarium completely and carefully; I have examined the herbarium given by Humboldt,—not complete but said to be as large as Kunth's own set or more so, and labeled by Kunth; I have looked at everything here which I thought could interest us, but some I found not, such as Cercocarpus; I have examined some other separate sets of the same kind. I am now ready to glance through Jussieu's herbarium, which is said to contain many Lamarck and Poiret; to spend a little time in Richard's, a few hours more for Desfontaines at Webb's, and perhaps Berlandier's[86] plants, though these are distributed through Webb's immense collection; this I can do, however, in evenings. Then a morning or two at Delessert's, which will be more occupied with examination of books than plants, will, I believe, finish. Webb has promised to give me some plants of Labilliardière, whose herbarium he bought, as he did Mercier's, in which he got many of Nuttall's plants. He has also a collection of Lady Dalhousie's from North America, all Drummond's, etc., etc.; so he is pretty rich in North American plants, but they are not all arranged yet. Webb has most generously presented me with a complete copy of L'Héritier's Works (in sheets) except the "Cornus," which I have this day bought of the Jew Meilhac, and for which I was obliged to give six francs. I shall have the whole bound in two large folio volumes: "Cornus" and "Sertum Anglicum" in one, "Stirpes Novæ" and "Geraniologia" in the other. I think thus far that the few copies of the "Flora" I have given away have turned to good account. I meant to go to Jussieu to-morrow, but Webb has made an appointment with me to see Dr. Montagne (muscologist, etc.) and his microscope, which is one of the latest and best of Chevalier, and will enable me to decide if I may venture upon one for Sullivant.

On Saturday Decaisne told me, almost by accident, that he was to do the Asclepiadeæ for De Candolle's "Prodromus," at the same time showing me a paper of his on the family that I was unacquainted with, much to his surprise, but he at once gave me a copy. You must know, that although I knew nothing scarcely of this family when I left you, and now know little as to general structure, yet I pride myself a little on my researches in extricating the synonymy of the species in London, in Herbarium Linnæus, Hort. Clift., Herbarium Gronovius, Banks, Walter and Pursh, and here of Michaux. Accordingly on Monday (yesterday) Decaisne and myself had a regular examination of all the species we could find here, and I furnished him with all my notes upon the synonymy, and left with him those I had with me from your herbarium, to be returned to London in September next. Decaisne has been with me also all this evening.

I find that very many of the pamphlets we have sent from time to time have miscarried, particularly the copies of my "Ceratophyllaceæ," sent by Castilneaux, and, what is mortifying, Guillemin and Jussieu received copies, but Brongniart and Decaisne none. I have just sent my only remaining copy here (for you sent me none) to Brongniart,[87] with an explanation.

There is a second species of Podophyllum from Cashmere or Himalaya, P. Emodi, also collected by Jacquemont, from whose specimens Decaisne has given me a piece. What is most curious, it is sixandrous, and therefore comes into Berberideæ except in wanting the dehiscence of the anthers by valves (which Decaisne tells me is also the case in Nandina), and so Robert Brown's views are confirmed. I should not wonder if the sly old chap had seen a specimen from Wallich when he appended the note to the "Congo Voyage" on Berberideæ.

Thursday evening, April 4.—Yesterday saw Dr. Montagne, the muscologist, and examined his microscope thoroughly, which is one of the latest and best of Charles Chevalier's. To-day I spent the morning at Jussieu's, looking up Lamarckian species, etc., in A. L. de Jussieu's herbarium; was very successful in Hypericum, but have no time now to give you details. In the afternoon Webb, by appointment, met me at the Garden, and we went to see Mirbel,—a man well worth seeing, I assure you. Webb acted as interpreter, when it was necessary, for Mirbel speaks with such distinctness that knowing what he was about I could understand him pretty well.

I like Mirbel excessively. Considering I was a perfect stranger, of whom he knew nothing, I think he took great pains to show me what I wanted to see. Sullivant's microscope will be of the same kind as his, only better, so that he will have the means of being a second Mirbel. Examined his microscope, which is a good one, but I think not equal to the best English; got some good hints, etc.; am to call again. He is very communicative, and you missed much in not seeing so extraordinary a man. He showed me a series of drawings and engravings on which he has been long engaged, for a mémoire on the structure of roots,—splendid drawings; and he explained to me what I before could not form a clear idea about, how the curious emboîtement or thickening of the walls of cells takes place by the development of new cells within the old. He showed me what I at once recognized as the so-called gridiron-tissue which I had seen in England, and I noticed that he explained it in the same way as Brown. He promised me copies for self and friends of the late paper of his on Embryologia in the "Comptes Rendus," just now read before the Institute (which will also be published with a part of the plates in the "Annales des Sciences Naturelles" and finally completely in "Archives du Muséum "), in which he says he has completely upset the new-fangled notions of Schleiden, Unger, etc. (adopted by Endlicher); and, what is remarkable, his investigations on the subject were made before he knew of their views, and the publication is only a little hastened on account of theirs. This evening I have been with Webb, looking up Desfontaines and Poiret plants, also some of Spach. Did I tell you I have seen a good deal of Spach of late? He does not agree well with the other botanists of the Garden; but there are some good points about him, and he is mending every day. I pushed him rather hard upon some of his bad ways, particularly that of his changing specific names, which he bore very well. Webb says he is now falling into an opposite extreme as to species, and will hardly admit anything to be distinct; but Webb himself rather inclines to multiply species, I believe. I am to meet Spach at his place in the Garden to-morrow morning. He is married, lately, to Miss Legendre, a relative of Mirbel's, who made his drawings in Marchantia, etc.,—indeed the best botanical artiste in Paris. What a fine library Jussieu has! And what a capital advantage it is to have a great botanist for one's father! I particularly envy Jussieu his collection of botanical pamphlets, which fill a large cabinet, all arranged in families, etc., the largest collection of the kind in the world, Jussieu thinks. He gave me to-day a little print of his father taken in the year his "Genera Plantarum" was published. He told me, what I did not know before, that Bernard de

Jussieu superintended the publication of Aublet's "Plantes de la Guiane." I could buy that work rather cheap, but think I must refrain. I bought to-day Schreber's edition of the "Genera Plantarum," two francs, two vols. in one, bound, for myself (you have it, I believe), and a second copy of "Linnæi Species Plantarum," ed. 3 (which is the 2d Holm., as you know, reprinted paginatim at Vienna). I gave five francs, and shall put it down for Sullivant, who should have it, unless indeed you desire to keep it yourself. I have bought (ten francs) the first four vols. of "Mémoires de l'Institut," 4to, bound, for library of Michigan. Ventenat's mémoire of Tilia is contained in one, also other botanical papers, and some good old chemical ones, etc. Webb is to put up for me a small parcel of Labilliardière's New Holland plants.

I have bought L'Héritier's "Cornus," so now I have the whole complete, and must get it all bound.

P. S.—I have just discovered that the copy of L'Héritier's is imperfect. I feel confident that Webb knows it not, and I of course cannot tell him. I shall have all bound up in one thick volume.

Monday evening, April 8.—I finished early this morning, at Richard's, the examination of those species upon which Michaux's herbarium is not satisfactory. Richard boasts of his set as the authentic one (which is true), but it is not as complete nor as good as the other, which is partly owing to Richard having divided with Kunth when he could. Michaux must have made a capital collection, since it has moreover supplied the general herbarium with a pretty extensive set, and Desfontaines and Jussieu with many; others I meet in the Ventenat herbarium (Delessert). They say De Candolle has some of Michaux's plants, and who besides I know not....

But I have something better than all this to tell you. I have discovered a new genus in Michaux's herbarium—at the end, among plantæ ignotæ. It is from that great unknown region, the high mountains of North Carolina. We have the fruit, with the persistent calyx and style, but no flowers, and a guess that I made about its affinities has been amply borne out on examination by Decaisne and myself. It is allied to Galax, but "un très-distinct genus," having axillary one-flowered scapes (the flower large) and a style like that of a Pyrola, long and declined. Indeed I hope it will settle the riddle about the family of Galax, and prove Richard to be right when he says Ordo Ericarcum. I claim the right of a discoverer to affix the name. So I say, as this is a good North American genus and comes from near Kentucky, it shall be christened Shortia, to which we will stand as godfathers. So Shortia galacifolia, Torr. and Gr., it shall be. I beg you to inform Dr. Short, and to say that we will lay upon him no greater penalty than this necessary thing,—that he make a pilgrimage to the mountains of

Carolina this coming summer and procure the flowers. Please lay an injunction upon Nuttall, that he publish no other Shortia, and I will do the same to Hooker in a letter that I am now writing. Indeed I think I will tell him some of its chief peculiarities, and then give him leave to publish the extract in the "Annals of Natural History" if he thinks it worth while.[88]

I attended a meeting of the Institute this afternoon. An election of a correspondent took place, which ran very close between Charles Buonaparte and Agassiz, but the latter carried it!

I must not forget to tell you about the Loganiaceous plant from Florida, for so Decaisne, to whom I gave leave to sacrifice a flower for drawing, has determined it to be; so Brown's hint is confirmed. There is something rather queer about the style, which, as Brown's "Prodromus" is not before me, I cannot say is also the case in any of the subgenera or genera he has indicated.

Euploca, Decaisne says, is certainly apocyneous. Nuttall, I believe, places it in Boragineæ.

April 9.—I heard Mirbel lecture to-day, commencing his course at the Sorbonne. He is a very good and clear lecturer, of the colloquial sort, and illustrates very well by rapid sketches on the blackboard. I believe you did not see him. In the contour of his features and in expression he is a good deal like Dr. Peters, except that his countenance is more attenuated, his features small and very little prominent, and his complexion light. At the Ecole de Médecine I was not fortunate enough to hit the chemical professor. I heard a portion of a lecture in the anatomical theatre, but soon came away.

I have had another fine lesson from Mirbel. He showed me all the drawings of the paper, of which I send three copies. I quit to-day.

LYONS, Wednesday evening, April 17, 1839.

At six o'clock precisely the malle-postes for every part of France began to leave, one after the other: that for Lyons came up; our baggage all in, our seats selected and arranged for us, in ten seconds we were in our places, and before the word adieu was fairly beyond my lips we were off at full speed. We took the route by Burgundy, passed Sens in the night, breakfasted at six next morning at Auxerre, and during the day should have passed through Autun, but I believe we did not; passed Châlons-sur-Saône at dusk, and arrived at Lyons at six precisely the next morning,—a rather fatiguing ride, but I saved much time over the diligence, which would have been even more fatiguing. The mail-coach takes four passengers only, three inside and one with the conducteur; it is drawn by seven horses guided by a postillion, in boots almost as high as himself, and the horses are changed every five miles or thereabouts. The time it took to change the horses I believe never exceeded a minute. I timed them once or twice by the watch, and we were moving again before the expiration of the minute. The country through which we passed was more fertile and in better cultivation than what I saw of Normandy; it was beautiful but monotonous, except the latter part, which grew quite picturesque as we approached the Rhone and the rivers that fall into it....

Lyons is finely situated just above the confluence of the Saône and the Rhone, occupying the space between the two rivers and also the other bank of the former. It has two beautiful and very steep hills, between which the Saône winds, which add much to its appearance....

April 25.—I broke off here some time ago, and left a space which I intended to fill up the first spare moment, by telling you what I saw at Lyons; what kind of a town it is; how I might possibly have seen Mont Blanc from it had it not been a rainy day; how I called on Seringe,[89] saw the little botanical garden, took notice of many little contrivances, particularly the way he keeps the aquatic plants wet; how he went with me to the Académie of Lyons, the branch of the University of Paris.... I could also describe the manufacture of velvet, which I also saw, but for all these things time does not permit; a good opportunity of sending to New York occurring to-morrow morning. So I must leave the hiatus....

I was called this morning at a quarter before four; went down to the steamboat, which was to start promptly at five, but which did not until half an hour later,—a narrow comfortless vessel, with no awning or protection for the decks, in which point, and in the lack of all comfortable arrangements, it is just like every other steamboat I have seen since I left New York, those between Liverpool and Glasgow alone excepted. The Rhone, even at Lyons and far below, merits pretty well the epithets applied to it, where it "leaves the bosom of its nursing lake,"—"the blue rushing of the arrowy Rhone," for it is rapid the whole course. At Lyons it has a blue tint like that of the ocean, though not so deep. Well, we were off at length, and aided by the current we made very satisfactory progress. The distance by post between Lyons and Avignon is one hundred and sixty-seven miles, but including all the turnings of the river it must be much more; however, at six o'clock and a quarter the spires and battlements of Avignon, lighted by the setting sun, were in sight, and a beautiful sight they were as we drew near. The wall of the city, built by Pope Innocent VI. in the twelfth century, is still perfect, and very pretty, the architecture being what I should have thought. Moorish (judging from pictures merely); the numerous spires of this very ecclesiastical town rising above it; the huge rocky elevation next the river,—the site of the ancient fortress, and of old temples, churches, etc.,—and not least the ruined bridge of very ancient date, that still throws its beautiful arches half across the river, the lovely Italian landscape around, so fresh and green, the distant mountains encircling the whole, made it altogether as delightful a scene as one could wish to behold. But you must know that I am now in the region of the olive and myrtle, and have in the short space of three days concentrated, as it were, the pleasure we experience in watching the gradual approach of summer. The season is said to be later than usual at Paris; it is like April in New York,—a few warm days, but the evenings all chilly and most of the days raw and unpleasant, The horse-chestnut trees of the Tuileries were just bursting their buds; but every hour since, and particularly to-day, I have noticed little by little the advance. Here nearly all the trees have assumed their foliage,—that pure and delicate vernal foliage which we always so much admire, but which you enjoy very much to come upon in the way I have done, instead of waiting week after week, with every now and then a snowstorm, just to keep winter in remembrance. But I must not forget that I have seen snow also to-day. The summit of Mont Ventoux, which we have had in full sight since twelve o'clock, is covered with snow, its brilliant whiteness contrasting finely with the craggy brown mountains of lesser elevation, as with the green fields and tender foliage of the valleys. There is nothing very grand in the scenery of the Rhone from Lyons to this place. The upper portion is very much like the Hudson between New York and the Highlands, but I think scarcely as fine, if you make due allowance for the effect of the old villages, etc. (not half so comfortable as ours surely, but much better adapted to improve the beauty of the landscape), with now and then a gray ruin, which is a vast improvement. But from Tournon quite to Avignon, the scenery quite surpasses the Hudson, and exhibits such variety, moreover, that you are charmed continually: now bold and magnificent even; again, picturesque, particularly where the basaltic rocks, for it is wholly a volcanic country, form parapets like the Palisades, but much more curious and diversified, the more friable material being worn away in places, leaving columns and salient portions in all fantastic shapes. And again, especially in the lower portion, we see the hills widely separated, leaving most beautiful broad valleys between, with high mountains for a distant background. At St. Esprit we passed under the curious old bridge built in the eleventh century, which is still in as perfect a state apparently as if finished but yesterday. It is three thousand feet long, and is said to be the longest bridge in Europe; it consists of twenty-six arches, and each abutment has also a little arch above it. We passed other very pretty or striking views of which I should like vastly to have good prints, but I do not know whether any person has of late been illustrating the Rhone. But I must come to a close, not to fatigue you longer. I arrived at the most excellent Hôtel du Palais Royal (recommended by Bentham) just in time for the table d'hôte at seven o'clock, and after dinner sallied out, with a guide to conduct me to see

Requien,[90] to whom Bentham had given me a letter. I found him a prompt man, and in almost ten words we settled my plan for to-morrow, which is to start in a cabriolet for Vaucluse at five o'clock in the morning, arrive at eight, spend two hours, breakfast, and return here by one o'clock; spend the afternoon and evening in seeing the most interesting objects in town, looking at his collections, his pictures, etc., etc. What would you give to see Vaucluse? I have many doubts whether it will equal my expectations, which are raised by the description; according to the account it must be very curious and strange, apart from the associations of the place, which here pass for little with me, as I feel no interest at all in Petrarch or Laura, whoever she may have been.

AVIGNON, Friday evening, April 19, half past eight o'clock.

I think you will scarcely call me an idle lad. It was about midnight when I went to bed last night; I was called this morning at half past four; a few minutes past five I was on my way in a cabriolet for Vaucluse, with a very lazy horse, so that it was nine o'clock when I arrived. I visited the famous fountain, admired the rocks, etc.; collected a few plants as a souvenir; took my breakfast, a very substantial one, consisting in part of delicate trout from the stream which issues from the fountain; left at eleven, arrived at Avignon again at half past two; saw the Requien museum of antiquities, which is rich, the paintings, the little botanic garden; saw also Requien's library and collection of plants, etc; made arrangements for correspondence; climbed the rocky hill which overlooks the town and river; enjoyed the view; visited the cathedral (a small affair) which stands upon it; saw the old papal palace, now converted into a prison; returned to the Hôtel Palais Royal, and a most excellent hotel it is, which I hope you will patronize the first time you come to Avignon; dined at seven, having first secured a place in the diligence for Nîmes at ten o'clock this evening, where I hope to arrive by daylight and be ready to go on the same day to Montpellier, where I prefer to pass the Sabbath. Now I think this is doing pretty well....

MONTPELLIER, Saturday evening, April 20, 1839.

At twelve o'clock I left Nîmes; rode through a highly fertile and level country, mostly occupied with vineyards, getting now and then a distant view of the mountains of Cevennes on the right, and soon of the Pic San Loup, by which I knew we were not very far from Montpellier. At this last place we arrived at five o'clock precisely, and here I am quartered at the most comfortable hotel imaginable, the Hôtel du Midi. All my stopping-places being indicated to me by Bentham, I have no difficulty in choosing where to stop. Here you are not put into a little seven by nine chamber up five pairs of stairs, as is the inevitable lot of a single man traveling in the

United States, but I have a room like a large parlor, airy, the two windows looking into a pretty shady garden, a sofa, cushioned chairs, and every convenience you can think of. The town itself has nothing pleasant except its situation, but there are in it two delightful spots, which I sought at once, after having taken my dinner,—the Esplanade, very near me, an elevated plateau planted with trees, from which you have an extensive view of the country around. From this I had my first view of the Mediterranean, distant, I suppose, about eight miles! At the opposite side of the town is the Place du Peyrou, one of the finest squares in the world, on a fine elevation, descending by bold terraces into the country around, the green fields coming up on one side close to the parapet. The view is beautiful and very extensive, the Mediterranean on one side, the Pic San Loup and the mountains of Cevennes on the other, while toward the south, it is said, the Pyrenees may be seen in very clear weather. From this point I discovered the Botanic Garden, the oldest in Europe and in many respects still the finest. So I descended, sought out Delile the director, who it seems expected me, and expressed his delight in a most exaggerated and truly French manner. I stayed with him until nine o'clock; returned here, commenced this, but being fatigued soon gave it up and went to bed.

Monday morning, April 22.—Nearly all of the foregoing has been written this morning; but I cannot stay longer, as I should be stirring. There are many Protestants in Montpellier, it is said, but I fancy that they are chiefly not very pious, and as I should not understand the language well enough to be benefited, I thought it better to spend the Sabbath by myself. This was my first Sabbath on land in which I have not attended divine worship conducted in the English language.

Tuesday morning, April 23.—As early as possible in the morning yesterday I called on Lady Bentham, the mother of my good friend who has taken so much pains to aid me and her daughter, Madame Duchesnil; they live quite retired, and are occupied in directing the education of the son of Madame Duchesnil, a fine lad of about thirteen.... The ladies received me with great cordiality. I prolonged my call to an hour, and accepted an invitation to take tea with them this evening.... I went to the Garden, called upon M. Dunal,[91] the best botanist here, who, having lived single to the age of I should say fifty years, has found out that it is not good to be alone, and has just taken a wife. I did not stay very long, as I found when I called that he was not in his study, but I suppose in his drawing-room, and I could not be so cruel as to keep him from the company of his beloved.

I called next upon Delile,[92] but as he was not in, I spent a long time in looking over the Garden, noticing all the little details and arrangements that it would be useful for me to know. On his return we spent the remainder of the afternoon in looking over his plants collected in America.

I dined with him at six o'clock, and spent nearly all the evening.... They have not water enough, however, to supply the Botanic Garden sufficiently, which has a very barren soil, and in this dry climate, where it seldom rains from this time till October, it suffers greatly. The first view of this garden is very striking, but upon a more careful observation I see less to admire. Still I learn some thing from every garden I visit.

Previously to calling on Lady Bentham I had accepted an invitation to dine this evening with Captain Gordon, a retired officer of the British army residing here, a friend of the Bentham family, who, hearing from Lady Bentham and Delile that I was soon expected here, called par hasard at the Hôtel du Midi, to request that they would send him word when I arrived. On finding me he insisted on my dining with him this evening. I have this moment, while I was writing, received a note from Lady Bentham, asking me to call on her this morning, saying she has a collection of plants made by herself for her son George at some interesting locality among the mountains, a set of which she is to have ready for me, knowing, as she says, that George would surely offer them to me. Although I had arranged my time a little differently, of course I shall call immediately after breakfast. Lady B., who is now very aged, is evidently a very superior woman; she is a very good botanist also, therefore, as I do not know the plants of the south of Europe very well, I am a little afraid of her.

Marseilles, April 25, Thursday evening.—I broke off my narrative on Tuesday morning, two days ago. I must continue my brief account, and then close my letters to send from this port. After breakfast, Captain Gordon called on me, and we went together to Lady Bentham. We found his dinner hour so late that we were obliged to give up the expectation of returning to take tea with the ladies here. Delile joined us, and soon after I went with him to see the museum of painting and sculpture, which, by a curious circumstance, is the richest in France, except that of Paris. There are not a few of originals of great masters; two or three Raphaels; as many of Salvator Rosa, Rubens, Poussin, Carlo Dolci, etc., many of which I know from engravings. We went next to the Medical School, which occupies the former palace of the archbishop, who was ousted at the time of the revolution. This is one of the oldest medical schools, and for a long time very celebrated. It is declining now; they have no professor of very great talent at present, except Lallemand. I was shown the gallery of portraits of the professors from the commencement almost, a prodigious number, and some of the old fellows very queer to look at. I saw also the library, the collection of manuscripts, classical, theological, a few Persian, Arabic, etc., which fell into their hands some years ago.

Thence we went to the Garden, looked at plants, but did not get on very much, Delile being fonder of telling long stories, complaining all the while how much he is pressed by his avocations, than of working hard. I then arranged my baggage, took a place in the diligence for Marseilles, called again on Lady Bentham, to take leave; dined with Captain Gordon, returned, and went to bed.

Rose on Wednesday (yesterday) morning at half past four; took diligence at five, arrived at Nîmes at half past ten; had time to take another survey of the Amphitheatre, the Maison Carrée, and so forth; took breakfast at half past eleven; off again at twelve, passed in sight of Beaucaire and Tarascon; crossed the Rhone, here a large river, near its mouth at Arles, a curious old town which has nothing modern about it, and thus was again in Provence. The court of Constantine the Great was for several years at Arles, which was celebrated for its refinement, and the women and children are said to be still handsome and graceful. Certainly nearly all I saw, young or old, were comely, and many handsome. They are all brunettes, and not a little sunburnt; but their black hair, large dark eyes, and long eyelashes appear to advantage. We were soon on the road again, traveled over an immense plain, bordered on the north by a long ridge of mountains, composed of naked jagged rocks,—a picturesque range, in fine contrast to the fertile plain from which it abruptly rises. They are, I believe, the mountains of the Durance. At length the plain became as barren as the mountains; night came on, and rather late in the evening we reached Aix, took our supper.... I slept pretty well, and when I awoke we were in sight of the town and bay of Marseilles, the latter superb as seen from the elevated place of our view; but the town did not present such an imposing view as I had been taught to expect....

Genoa, April 27, 1839. Saturday evening.—I have just finished my afternoon and evening stroll through this, to me, the first Italian city: the birthplace of Columbus, the city of the Dorias, the rival and even the conqueror of that other proud republic of the Middle Ages, Venice, in remembrance of which, huge pieces of the chains which were employed to bar the harbors of the latter city are suspended from the gates of Genoa. We arrived in the bay before twelve o'clock to-day, and during our gradual approach to the town enjoyed the view to the full; both the distant view and the near are very fine,—equal, I may say, to what I expected, which is saying a great deal. As seen from the bay it certainly deserves the name its citizens long ago gave it,—Genoa the Superb. You have the whole completely before you in one view, the buildings rising one behind the other, the fortifications that overtop the whole, with the vast mountain amphitheatre for a background.... You are not much disturbed with the rattling of carriage wheels here. With the exception of one street, and this a new one (Strada Nuova) at least as to its present dimensions, they are barely wide enough for a wheelbarrow, and mostly too steep for a carriage, even if they were wider. The houses are very high; six, seven, or eight stories being very common, indeed usual, so that the streets are mere chinks or crevices. I found the same advantage from this in Avignon and the other towns of the south of France, that is, the perfect protection afforded these warm days from the heat of the sun. You are sure of shade; and the air is so dry that none of the inconvenience and unhealthiness results which would surely be the case in other countries. I am at the Hôtel des Etrangers, not far from the quay, and my room, five or six stories high, looks down upon the harbor and bay. It is nine o'clock in the evening. The light is burning quietly in the light-house, a tall and very slender column at the entrance of the harbor, forming a beacon which is visible far and wide. I don't know as I may say that

"The scene is more beautiful far to my eye
Than if day in her pride had arrayed it;"

but it is much softer. The evening gun has just been fired off from one of the batteries next the sea, the signal, I suppose, for closing the harbor, and the echo sent back by the hills on either side was prolonged and repeated fainter and fainter for nearly a minute....

The coast at Marseilles and that I saw yesterday may be described in a few words: bare, jagged, sterile, rocky mountains; scarcely high enough to be picturesque, perfectly destitute of verdure, barely supporting here and there a few stunted olive-trees. We passed Toulon and had a distant view. We sailed between the mainland and the islands of Hyères, so remarkable for their fine climate and healthfulness, but they did not look very inviting to me.

When I rose this morning the scenery had become bolder and more interesting. We were where the Alps first come down to the sea, and we have since sailed along a coast so closely skirted by the Maritime Alps, the chain which passing into Italy forms the Apeninnes, that there is scarcely room to construct a road between. The loftier peaks, the whole day, were covered with snow, in fine contrast with the gray and sterile cliffs below and the dark blue sea which seems to lave their base, for the Mediterranean has the deep azure tint of mid-ocean quite up to the shore. There are many pretty villages also, which either seem hung on the mountain's side or to rise out of the water. In one place I counted twelve in a single view, by no means a wide one. We passed Savona, the town where the pope lived while Napoleon was master of Italy. Here the hills are more fertile, and vines, olives, and oranges are cultivated wherever room or soil enough to plant them can be found....

IN THE HARBOR OF LEGHORN, Monday evening, five o'clock.

I must tell you of the pretty view I had Saturday night. My room, I think I mentioned, looked directly into the bay, and also gave me a fine view of the western part of the town, the mountains of that side of the bay, and peeping over them, the sharp crests of the Maritime Alps, still white with snow, and looking rather like bright clouds than a portion of terra firma.

While I was sleeping soundly, about two o'clock in the morning the moon shone into the window directly into my face, and thinking it a pity I should lose so fine a sight, she awoke me. She was near her full; she hung in the middle of the bay at just the proper angle that the flood of golden light she was pouring upon the tranquil sea was reflected directly to my eyes. The city, too, looked beautiful indeed, and the mountains, and even the Alps, were all visible. I enjoyed it for a long time, and went to bed again regretting that I had no one to share the scene with me.[93]

There is or was a British chapel here, belonging to the British embassy, but I could find nothing of it, and so spent the Sabbath by myself, which was as well perhaps. At seven in the evening our boat left, and I was obliged to continue my voyage. I wrapped myself in my cloak and slept soundly and quietly, and when we reached the harbor of Leghorn at five o'clock awoke refreshed, vigorous, and in the finest spirits. I obtained a light breakfast on board; at seven o'clock was ashore; in five minutes more was in a cabriolet and on the road to Pisa, distant from here fourteen Tuscan miles, which make, I should judge, about ten English ones. My bargain was that I should be driven to Pisa in two hours at farthest, have two hours and a half there, and be returned again safe and sound before two o'clock. This was easily accomplished; the journey being made in less than two hours, I had the more time there, quite as much, indeed, as I wished. It is a great comfort to be able to leave a place the moment you have done with it, and so avoid being sated with it. I had a letter and a little parcel from Mirbel to deliver to old Savi,[94] the professor of botany in the university; so I was dropped at the door of the university, once so famous, but now far from formidable. I found Savi, gave my letter, was introduced to his two sons, the one professor of natural history, the other assistant professor of botany, who showed me through the museum, which was interesting, the botanic garden, which was not much; I then set out to see the four chief lions, the Duomo or cathedral, the Baptistery, the Campanile or famous leaning tower, and the Campo Santo, which all stand near each other and are soon dispatched. In fact they are the separate parts of a cathedral, the Campanile being, as the name denotes, the bell-tower, and the Campo Santo the burial-place....

The vine in Tuscany is not kept close to the ground as in France, but is trained in arbors and festoons along the borders of wheat-fields, and when their leaves appear must add very much to the beauty of the country. One here could sit under the shade of his vine, which would be out of the question in France. But the boat is leaving the harbor. On the right we can dimly discern the northern extremity of Corsica. Elba we shall pass in the night, and sometime in the course of the morning be landed in Civita Vecchia. I have made the acquaintance of an English clergyman of warm piety, who is in ill health, who has been obliged to reside for several years in Nice in the winter, and at Interlaken in Switzerland in the summer, at both of which places he preaches regularly. He has traveled in Greece, Turkey, and Asia Minor, and passed much time with our missionaries there, of whom he speaks in the warmest terms. His name is Hartley. We shall go on in company to Rome.

ROME, 1st May, 1839, Wednesday evening.

And I am indeed in Rome. This is enough to repay one for long and tedious journeys and even for transient separation from friends, and when I leave this place I feel as though my face was set homeward. I feel it is something to be in Rome....

I distinctly recollect the time when, a very small boy, in the course of a long ride with a relative, the story of Romulus and Remus was first related to me, and how it struck my wondering fancy. And I recollect most perfectly my first lesson in Virgil, and how, commencing with "Arma virumque cano," I slowly worked my way into the mysteries of Latin prosody and the story of the Æneid. Little did I think in those days that I should ever stand within the "walls of lofty Rome;"

"Should tread the Appian
Or climb the Palatine, and stand within those very walls
Where Virgil read aloud his tale divine."

My enthusiasm has risen by degrees, for I arrived here this morning, after a delay at that most wretched of all places, Civita Vecchia, where an Austrian soldier, stationed there, told us he was sent as to a kind of earthly purgatory to do penance for his sins; after being subjected to those numberless petty exactions by which the purse of the pope is replenished from the pockets of us poor Protestants, after tedious delays on the road, and a most uncomfortable ride for the whole night, which altogether is enough to put one in a bad humor with everything,—after all this you may be sure I found myself in such a prosaic care-for-nothing mood that it was a long time before I could feel the interest which the Eternal City is calculated to inspire. A fog in the morning prevented us from a good view on our approach; the streets of the modern town through which we passed were mostly devoid of interest, and we saw nothing but the dome of St.

Peter's and the Castle of St. Angelo. However, we got established at the Hôtel d'Allemagne, and took breakfast. Mr. Hartley, being worn out by the journey, took to his room for the day, and I was left to myself. Though perfectly ignorant of localities here, I was determined not to be deprived of the satisfaction of discovering the most interesting places for myself. My guide-book (Madame Starke) describes objects somewhat particularly, but gives no information as to where they are to be found. I hate the chatter of a cicerone, and felt confident that I should stumble upon something worth seeing. So I climbed the hill just before me by a magnificent flight of marble steps, where the Egyptian obelisk stands which the inscription says was found in the Circus of Sallust. I saw an imposing building at the end of a long avenue, on the summit of a rise which I afterwards learned was the Esquiline Hill. On reaching it and examining the interior I found by the guide-book that it was the Basilica of Santa Maria Maggiore. These basilicas, retaining the name of ancient structures, are a larger kind of churches, which were mostly established upon the foundations of ancient temples, or they were these temples themselves turned into churches....

As I emerged from the Coliseum I stood between the Palatine and the Cælian Hills, the Arch of Constantine just before me, the Arch of Titus in view on the right hand, and just beyond the Roman Forum, all crowded with ruins; the very soil is mouldering brickwork and fragments of columns. Here I spent the greater part of the morning, silent and undisturbed, finding out by the description the ruins as they presented themselves....

The journal is so long that most of the Italian, more especially the Roman, journey must be omitted. Dr. Gray, as is shown, was a busy sightseer, enjoying the historical and romantic associations with his natural enthusiasm. Here began his great love of painting, of sculpture, and of architecture; he carried the details of churches and cathedrals in his memory remarkably, recognizing quickly a print or photograph of something he had seen perhaps thirty years before; he had the memory for form which helped him so much in his science. He was a good critic of painting and enjoyed extremely his favorite pictures, liking to wander off alone to enjoy them. Titian on the whole ranked highest in his estimation. He enjoyed much of the old church music, though his preference in music was for simple songs, hymns especially, and the old tunes to which words had long been wedded. There are many quotations from Byron and Rogers in the original journal. For Byron, with his brilliant descriptions and versification, he always kept much feeling; and his great love of natural scenery had full play.

TO MRS. TORREY.

LEGHORN, May 8.

Whenever I have an hour to spare I know of no pleasanter mode of occupying it than by writing to you, for to you my thoughts, whenever they are at rest, spontaneously revert. I have yet an hour before the vetturino starts for Florence, and I may as well commence another sheet, the first of a series which I may be unable to send you for several weeks, as I here leave the Mediterranean, loveliest of seas, and except I find an American ship on the Adriatic, which is not very probable, I must keep them all until I reach Hamburg. I have just closed a formidable packet of journal, to be sent from here in the ship Sarah and Arsilia, which is to sail for New York next week....

I am very well satisfied with my visit to Rome. In the brief space of time I spent there I saw everything I wished except the pope himself, and I believe I had a glimpse of him; one statue of Michael Angelo's, which I only learned about when it was too late; the Catacombs, where the early Christians used to conceal themselves, which are some miles off; the monument of Cecilia Metella, which is not handsome, but is immortalized by three or four singularly sweet stanzas in "Childe Harold;" and the Basilica of St. Paul, which is some distance out of the city, and was nearly destroyed by fire about ten years ago. This is a very small list compared with what I have seen, so I am quite content. I wish you could see Rome; there is so much that you would enjoy in the highest degree, and it is laying up a fund to be enjoyed afterwards as long as you live.

It is now just sunset, and the air is remarkably balmy,—a mild sea-breeze, just enough to fan you. And let me tell you, however, as to Italian skies and sunsets that they are not a bit superior to our own. You may enjoy from your own parlor windows finer sunsets every clear day in summer than I have yet seen in Italy; though they certainly are very near ours. It is only to those who are accustomed to British clouds and fogs that they are remarkable.

The peripatetic grinders of music upon hand-organs so common in all our towns are usually Italians, and I supposed that street music here was of much the same kind. This is a mistake. I have not seen such a thing in Italy or the south of France. You have universally the harp, commonly two players in concert, and very frequently a violin also for accompaniment, and the music is always creditable. At Avignon, the very land of troubadours, we were serenaded at dinner with a concert of harps, guitars, etc., but when they called for the coppers we found, shame to this degenerate age, that the troubadours were all women, and of the most unromantic appearance possible. The patois of all this part of France and of Piedmont, however, is the same as the language in which the trouvères are written, and one who understands the patois as now spoken can read the former without difficulty.

The Italian language, is very soft and musical, far more pleasant to the ear than the deep nasal tones of the French.

JOURNAL.

FLORENCE, May 9, Thursday evening.

Finding little more that I could do to-day, I then called at the residence of Mr. Sloane, a descendant of Sir Hans Sloane of famous memory, who resides in the Bontrouline palace, and not finding him at home left a note of introduction written by two ladies, Mrs. Boott and Miss Boott, and also a letter intrusted to my care by Mirbel. I called also at the Botanic Garden, but Mr. Targioni-Tozzetti[95] was not at home, and the garden was of no great consequence. While at dinner Mr. Sloane called to welcome me to Florence, and to take me out of the city to the Campagna,—lawns and beautiful pleasure-grounds and groves skirting the Arno for a mile or two, which are thrown open to the public, forming the favorite drive or promenade. Almost the whole city was there, and I never saw a more pleasant place. The roads were thronged with carriages, from the barouche of the grand duke to the peasant's cart, all on terms of perfect equality. The grand duke passed us twice. He mingles much with the people, is accessible to all, and is greatly beloved. The government, though despotic, is paternal, the people are not burdened with taxes, and are contented and industrious. The difference between Tuscany and the Papal States is manifest enough. But I must hasten with my narrative. Early the next morning, Friday, I called on Mr. Sloane, looked at his garden, where he has many fine things. We then crossed the Arno to the other side of the town, called on Professor Amici,[96] who removed here from Modena a few years since, and has charge of the grand duke's observatory. He was very obliging, showed me his microscopes, which he thinks unrivaled, but I don't, and then the observatory, where I saw all the instruments, peeped through his telescope, and from the top of the tower had a most beautiful panoramic view of Florence and the surrounding country. We then passed through the museum of natural history, which is in the same building, and is prettily arranged; saw the famous flowers and fruits done in wax, but not the figures which represent the Plague, which were in the anatomical museum adjoining, and which I did not care to see. In the collection were some recent models made under Amici's superintendence to illustrate his discoveries, etc. They were wonderfully fine, and would be useful in a class-room. Amici is a good observer with the microscope, but his anatomical or physiological notions are in some cases very wide of the mark, and quite surprised me.

On leaving, Mr. Sloane and myself separated, he going to fulfill some engagement, and I to the Palazzo Pitti, as it is still called from the founder, though it early passed into the hands of the Medici family, who finished it, and now it is the ducal residence. I must tell you, by the way, that I should have seen a remarkable person in Florence, had she not been sick. Sloane is very intimate with her and wished me to see her; she is the ex-queen of Naples, the widow of Murat and the sister of Napoleon....

On returning to the hotel, however, I learned that I could not get a place with the courier next day, that the diligence which left at mid-day did not arrive at Bologna until Sunday afternoon, so I engaged a cabriolet, to start with me after dinner, arranged my affairs, called on Mr. Sloane to bid him an unexpected adieu, dined at the table d'hôte at five, and at dark I was climbing the outskirts of the Apennines.

I would have liked to call upon our sculptor Greenough[97] to see how the statue of Washington is coming on, but had not time.

At sunrise I was on the mountain-summits, among the clouds, which a strong wind for a moment blew aside, and gave me some magnificent views. We journeyed for some hours in this elevated region, but at length crossed the Tuscan frontier and were once more in the country of his Holiness. Just as we commenced our descent, which is very abrupt, a dense fog enveloped us and it began to rain; in consequence of this I lost the view which you often have of the Adriatic and the Mediterranean at the same time, as well as the plains of the Po on the north. This was the first rain I encountered, excepting a few drops at Rome, since I left Lyons; so you may judge of the dryness of the climate in the south of France and Italy. It is very different, however, near the mountains. At length, after a long and rapid descent, we arrived at the foot of the mountain, and stopped at a comfortable inn to take our dinner and breakfast at once, it being about two o'clock. Several carriages were there before us, and just before I left another arrived, bringing with it a most genuine Yankee, who amused me excessively. It seems that he came out in the Great Western, a few weeks ago, had seen what he thought worth seeing in London and Paris, had been even to Naples, and was now on his way from Rome to Switzerland, and expected to reach London to return by steamship in—I forget how many days! But the feat upon which he prided himself above all was that he had ascended Vesuvius and come back again in—I don't remember precisely how many minutes, but in an inconceivably short space of time, and very much quicker than had ever been done before! to the great wonderment of the guides, as he said, and as I do not doubt. This was his chef d'œuvre, and I assure you he felt quite proud of it. I laughed most heartily at the absurdity of the thing, until I reflected how rapidly I had been doing the sights myself, and felt I might justly come in for a share of the ridicule. In this day's journey I think I outdid the Yankee, for, arriving at Bologna about five o'clock, I immediately made arrangements for going on to

Ferrara the same night, and this accomplished, I had but two or three hours to spend at Bologna, a city famous for its university and its sausages; the former decayed almost to nothing, the latter still in great demand, diffusing their abominable garlic odor from every table. I visited all the large churches, took some coffee, and before nine o'clock was on my way through the vast plain watered by the Po, which, like most large rivers, branches near its mouth into several streams. The lad who drove me did not know the road very well, and lost his way several times, so that instead of arriving before daybreak it was six o'clock in the morning when we entered Ferrara. Indeed he came near losing his horse as well as the road, for while I was sleeping soundly in the carriage I was roused by a prodigious clatter, and jumping out as quick as I could, found that he had driven into a heap of rough stones deposited to mend the road; the horse had slipped and was lying flat upon his back in the bottom of the ditch. With much ado we liberated him from the carriage and lifted him out of the ditch, repaired the injury to the harness as well as we could with bits of rope, and were again on our way. I have wondered since how I could ride thus through the night, with only a boy with me, through a country which some years ago would not have been deemed safe. But I felt not the slightest alarm, and slept as soundly as possible.

Ferrara is famous for possessing the tomb and chair of Ariosto, but except this is as uninteresting as you can imagine. It was Sunday, and I spent the day within doors as well as I could.

By making a very early ride I succeeded in reaching Padua at ten o'clock this morning; visited the university so famed of old, the churches, the splendid Caffè Pedrocchi, the Botanic Garden,—the most ancient in Italy, of which Alpinius, the elder and the younger, and Pontedera were the directors. It is under the care of Visiani,[98] to whom I brought a letter from Bentham, and who politely showed me all I wished to see. The university is a queer old place indeed, and the lecture-rooms the most dark, gloomy, and incommodious places you can conceive; everything is as old as the fifteenth century. I wish I could describe the anatomical theatre, which is the most curious specimen of antiquity I have seen. The Museum of Natural History is so-so. There is still a goodly number of students, but nothing to what there was in the olden time. The Duomo is a small affair, but the church of St. Antonio is like a mosque, the most Saracenic building I ever saw,—with its seven or eight balloon-shaped domes of various sizes, and three or four tall and slender minarets. I am sorry I can't get a decent print of it. The interior is noble, and very rich in tombs and shrines and sculptures. Here are tombs of many of the old professors. The church of St. Augustine is in the same style, and not much inferior.... There is very much that I wish to write, but I have not the time nor the strength to write longer, and must sleep. To understand the full luxury of a bed you should sleep without one, as I have done very often of late. Good-night.

VENICE, on board steamboat for Triest, lying at anchor,
Wednesday evening, May 15, 1839.

For nearly two days I have been "a looker-on in Venice," a strange place, as unlike any other city of Europe as can be, unless Constantinople resemble it in some respects. It is more like some place you visit in dreams, some creation of fancy, than a real, earthly city, if it can be called earthly which scarcely stands upon earth.

We left Padua at five o'clock in the morning, yesterday, by the diligence, passing along the banks of a canal, bordered with numerous villas; all of them had been fine, some very magnificent, but they are now decaying. The clouds prevented me from obtaining a view of the Rhætian Alps, which bound the view on the north, but I hope to make up for this to-morrow, which will give me some amends for our detention here; for you must know that the steamboat was to have left at nine o'clock this evening, and I expected to have been in Triest this morning; but the day has been stormy, and the water is a little rough, so, forsooth, the boat is to remain until morning; but as it is to start early, I have remained on board, where I have a comfortable place to sleep, and a quiet hour to write.

Oh, I wish you could see Venice!—and the dear girls—whenever I see anything particularly queer, I think of them at once, and wish for them to enjoy it with me. And here everything is strange, canals for streets, gondolas for coaches; not a horse to be seen in the city, except the celebrated bronze gilt steeds of St. Mark; palaces of barbaric magnificence, splendid churches; people of all nations and tongues, Christians, Turks, and Jews. Surely there is nothing like it. The view from Fusina, on the mainland, which was the first I obtained, was charming....

You will wonder at the comparison, but the distant view of Venice reminded me strongly of New York, as you approach from Amboy. The gondola that brought us stopped in the Grand Canal near the Rialto, or rather the bridge of the Rialto, for the name properly belongs to the island; and in crossing this bridge during the day, I found some of the little shops still occupied by money-changers, and I saw more than one hard Jewish countenance that might sit for the picture of Shylock. This part of the town is unpleasant, although the canals are lined with what were once stately palaces, which now look as if about to sink again into the water. While on my way to a hotel, I came abruptly upon a view that seemed like enchantment: the Piazza of St. Mark, a large quadrangle, three sides inclosed by a magnificent range like the Palais Royal; on the fourth, the church of St. Mark, and adjoining it the Palace of the Doges, scarcely less magnificent, and in an equally Oriental style. In front is the Campanile, taller than that of Florence, but not handsome. As you turn out of the quadrangle in full front of the palace, you see the two granite columns, one of them surmounted with the winged lion; and you stand on the mole, with the most superb view of sea and city, shipping, churches and palaces, before and around you. I never expect again to see anything like it. I have walked over this ground again; and one is never wearied with the sight.... The street musicians here are very good. A party stops at the door of the café: a man with a violin, his wife and son each with a guitar, and they perform several airs exceedingly well, the woman sometimes accompanying with her voice. She enters the café with the little wooden cup in her hand, and is well satisfied with a kreutzer (about half a cent) from those who choose to give, and a sweet "grazia" in the softest Italian expresses her thanks. There is one café here frequented almost exclusively by Turks, who sit smoking their large pipes with such an air of ridiculous gravity. Their turbans or the red caps they often wear, their flowing robes and their nether garments, which are something between pantaloons and petticoats, are very queer....

I spare you a detailed account of my movements to-day and yesterday, of the fine churches, enough to furnish cathedrals to half a dozen cities, of the arsenal, its ship-yard, the antique lions, the public garden, the Armenian convent, the gondolas and my rides therein. I have enjoyed it greatly, and have laid up a stock for future enjoyment, for I shall read hereafter of Venice with greater interest. One who travels as rapidly as I do, if he would enjoy the full benefit of his journey, should know almost everything before he leaves home. The true way for those who have time and means sufficient is to study the history of each place on the spot with all its monuments and relics around them. So more might be learned in one month than in a year at home. If I had what I am not likely to have,—a family of children to bring up, money sufficient for the purpose, and no other duties to prevent, I think I would educate them in this peripatetic way. But now to bed.

Thursday evening, May 16.... We are to start at nine o'clock. The rain is over, but it is still cloudy. I have been for some days in Austrian dominions, but I wish to be in Austria itself. It cleared up a little just at sunset, and gave, me from the deck of the vessel, a most beautiful view of the town and harbor, with hundreds of gondolas gliding swiftly through the water in every direction....

TRIEST, Saturday evening, May 18, 1839.

As misfortunes never come single, I found this morning that our places were not secured in the mail-coach for Monday. The fellow who was to arrange the business found, after getting our passports in order, that there was only one place left, and supposing that we were certainly to go together, did not secure that. It was immediately arranged between us that I was to have the place, but on arriving at the office I had the mortification to find it already taken. For an hour or so we made various plans, negotiated with a vetturino, but were stopped by the information we received, that they would be five days on the road to Gratz, from where to Vienna it would require at least two days more by the same kind of conveyance, or twenty-seven hours in the mail-coach if we could get a place in it. We found that the quickest way left for us was to take places for Tuesday by the mail, and go on Monday by a private conveyance to Adelsberg, as we had intended, where we shall have a day longer than we desire; and these places we were fortunate enough to secure. So I cannot expect to reach Vienna before Friday morning of next week! I had hoped to reach that place by the twentieth.

It rained hard all the morning, so that botanizing was out of the question. So I put my collection of yesterday in press; visited Biasoletto,[99] and after dinner met Tommasini,[100] who has given me a very pretty collection of plants of the country....

VIENNA, 24th May, Friday evening.

The great fête of the Grotto of Adelsberg, of which I wrote you, was to take place on Monday afternoon. Mr. Philip, the painter, and myself took a carriage to that place and arrived in good time, and saw this very strange grotto with greater advantage and under more curious circumstances, I suspect, than was ever done by an American before. I had all the next day before me, as the coach from Triest did not arrive till evening. My companion was taken somewhat ill and kept the house, while I took my portfolio and walked through the fields of this retired valley to a bold and high mountain range, more distant than I had calculated on; climbed the rocks with much difficulty; enjoyed a charming prospect from the summit; filled my portfolio with plants; got back about five o'clock, regularly tired and hungry, and just had time to eat my dinner and secure my specimens before the coach came from Triest. We took our places just at dusk, Tuesday evening, and have been on the road day and night, stopping just long enough to take our meals, until this morning; when at early daylight, just as I opened my eyes from such sleep as one might catch after three consecutive nights of such confinement, the vale of the Wien and the beautiful city of Vienna lay before me, the green fields reaching up to the very gates. It was a lovely sight. I have never seen the like. It began raining very soon, however, and has rained all day, so that I have seen little. Philip, who understands German, has been confined to his room by illness. But as soon as I got my breakfast and was fairly fixed in my lodgings, which we found as difficult to get as if we were at New York at this season (I am at the Gasthof zur Dreyfaltigkeit, a good and cheap house, and the head waiter speaks French), I took a guide to direct me to the Joseph-Platz, where the Imperial Library and Cabinet are, to find Endlicher.[101] I found the man in his den, and the moment I put my letters into his hand he recognized Bentham's writing and addressed me by name, Bentham having apprised him of my intended visit. Endlicher received me very cordially, and I remained with him till two o'clock. He is extremely good-looking, and younger even in appearance than I expected, although Bentham told me he was about his own age; he looks about thirty-three. I had the pleasure to present in person the copy of the "Flora" designed for him.

The usual dinner hour here is from twelve to three. The common people dine at twelve, the gentry from two to four, the imperial family setting a good example by dining between one and two. After dinner I went to the police office to procure the necessary leave to remain here for a week or so, answered all the questions which are put in such cases to the traveler, such as where I stopped, how long I intended to stay, what my business was, produced my letter of credit, in order to show that I was not likely to run away with unpaid bills,—to ascertain this point is said to be the chief object of all this inquiry. When you arrive at any hotel and remain over night, you are presented with a blank formula comprising still more particular inquiries, which you are required to fill up, and it is sent to the police office. You give first your name, then your country, age, religion, occupation, state whether you are married or not! whether you are traveling alone or in company; where you came from last; your probable stay; whether you have letters of credit or not, with some equally particular inquiries! I went next to my banker's, found no letters! I drew some money, and obtained a ticket of admission to a commercial reading-room, which is well supplied with English and French newspapers. Here I stayed until sunset, reading up my English news, in which I had got far behind, and which on the present occasion I found very interesting. I gleaned occasionally a little news from home, but vaguely. The information seemed in general satisfactory, but one letter from home were worth it all!

I have this morning changed the plants I have been drying, and have taken care of my companion Philip, who is quite sick with the fatigue of his journey and so forth. I have endured it very well, but must get into bed. Not having had my clothes off for three nights in succession, nor enjoyed rational sleep, I wonder much that I am not more fatigued. Endlicher asked me to go to the opera this evening, where there is some especially fine music, as he says, but I declined, telling him that under present circumstances I should sleep through the finest music in the world. I suppose it would be perfectly impossible to make him understand how one could have any scruples against this amusement.

Saturday, 25th, 1839.—I went early this morning to the Imperial Cabinet; remained there until two, when the rooms are closed. After dinner I explored about the city until sunset; saw many of the public buildings, the gardens, etc. I understand the localities of the town proper very well. The city itself is not large; the strong walls that inclose it are still kept up, and immediately outside of this there is a large open space, planted with trees and laid out into roads and walks. Beyond this are the faubourgs or suburbs, larger many times than the city itself; very pleasant, but rather inconvenient to reach. Most of the public buildings, the shops, etc., are in the city itself. I went to see the fine old Gothic Cathedral of St. Stephen's. It is a very old and exceedingly fine, large building, but the roof is very awkward. The spire is the finest thing I ever saw in the way of Gothic architecture. It is four hundred and sixty-five feet high, and is the very poetry of steeples. I intend to climb to the top presently....

Monday morning, 27th May.—I find we are in a different climate from Italy. It has been cold ever since my arrival here; the first day was rainy, and yesterday it rained from morning to night, and was very cold and unpleasant; so of course I kept my room nearly all day. I had also to take care of Mr. Philip, whose indisposition has turned into intermittent fever, such as he has been subject to at Rome. It is a most distressing thing to be sick in a strange land, and I cannot be too grateful for the uninterrupted good health I have enjoyed ever since I left you.

I have deferred telling you anything about the Grotto of Adelsberg, on account of the great difficulty I find in conveying any idea of it. It is without doubt the most wonderful thing of its kind in the world.

Adelsberg itself is a little German village perched under a steep conical hill which is crowned with the ruins of an old castle; it is at one border of a circular plain, several miles in extent, dotted here and there with little hamlets, and surrounded with mountains, so that it is like a large basin, and seems wholly shut out from the rest of the world. It is so still and quiet that it would do very well for the valley of Rasselas, but the mountains do not form precipices except on one side, where they are accessible at a few points only, and there with much difficulty, as I had occasion to know. The streams that come down from the mountains unite to form a little river, perhaps nearly twice the size of the Fishkill Creek; and this, after running about the valley seeking an outlet in vain, at length in despair, as it seems, dives into the solid rock at the foot of hills near the village. The entrance for visitors is a small hole above this, which opens into a long gallery, perhaps two hundred yards in extent. From this you descend into a vast hall, called the Dome, more than one hundred feet high, and three or four hundred feet in length. As you descend you hear the roar of the waters confined in their deep prison-house, and at the bottom you meet the river which rushes swiftly to the distant extremity of this hall, and there sinks into the dark depths. Instead of a stupid monument and inscription by the late emperor, placed above this, it would have been much better taste to have placed in the stream a piece of statuary representing Charon and his boat, for never was seen so perfect a beau-ideal of the fabled river Styx. This is the last you see of the river Poik; but the Unz, which bursts forth a large stream from the rocks at Planina, is believed to be the same. This river is crossed by a bridge. Then we went on to another hall about three quarters of a mile from the entrance; the ball-room, where a large gathering of peasants of the surrounding country, in their national costume, were dancing waltzes in the bowels of the earth!

Hiatus vastus.—I left this account of the Adelsberg Grotto, and my journey through Illyria and Styria, for the first convenient opportunity,—a time that never comes,—so now I must send it as it is. The grotto is wonderful past all description, and our visit was very opportune; the whole scene not soon to be forgotten.

29th May.—It rained all day yesterday, so Schönbrunn was out of the question, and I spent the morning again at the Cabinet of Botany; and after dinner Philip and myself, in spite of the rain, set out to visit the imperial picture-gallery in the Upper Belvedere Palace, which is finely situated in one of the suburbs. The gallery is very extensive and excellent, especially in the Dutch school, and we had barely time to finish our hasty reconnoissance before it closed for the night. I had a fine view of the city from the windows of the upper story. We stopped at a café on our way home, took some lemonade and ice-cream, while I read "Galignani's Messenger" for English news. This morning I went to the gallery as usual, and after working for a little time, Mr. Putterlich,[102] the sub-assistant, went with me to the famous Mineralogical Cabinet, the finest in the world. A most splendid affair it is. It occupies a suite of quite ordinary rooms, but is excellently arranged and shows to great advantage. Here are all the fine gems, diamonds, emeralds, topaz, and all sorts of precious stones, both polished and natural. I saw also the bouquet of precious stones made for Maria Theresa, a most brilliant affair. The collection of aerolites is unique. I intend to visit it again on Saturday. I obtained some useful information here as to the mode of constructing the shelves, etc., in a mineralogical cabinet; their plan here is the best I have seen. If I knew what I now do, I could have given a plan for the construction of the cabinets at the Lyceum infinitely better than the present. Returning to the Botanical Gallery I occupied myself in selecting specimens for myself from Rugel's New Holland collections. Endlicher offers me these and other plants, as many as I like. He also offered to send to Hamburg for me a copy of the

"Iconographia Generum Plantarum," the "Annals of the Vienna Museum," and some other of his works. After dinner, finding nothing else to do for a few moments, I went into a bookseller's,—the publisher of Endlicher's "Genera Plantarum,"—to look up some reports on education, etc. I asked also for botanical works; and after offering me several things which I did not want, they brought out, as a great rarity, our own "Flora," which I told them I did not want at all. At six o'clock, Endlicher called upon me to take me to the Botanic Garden of the university, under the care of Baron Jacquin, who is professor, at the same time, of both botany and chemistry in the university, and scarcely lectures on either. He introduced me to the old fellow, a hard-featured chap, who managed to speak a little English and talked to me of the year he spent at Sir Joseph Banks' in bygone times. We went through the garden, which is finely situated, covers much ground, and has fine trees, but is wretchedly cared for; in fact it is almost left to run wild, although well endowed.... I have some curious anecdotes to give you about the censorship of the press at Vienna, but have not energy enough left to write this evening.

Thursday evening.—Nothing can be printed and published here, without first being examined and approved by a censor of the press. The government appoints four or five persons in Vienna, who examine in different departments, one for newspapers, one for works of science! others for different branches of literature. Every author must send his manuscript to the police-office, whence it is handed over to the proper censor, who certifies that it contains nothing immoral, nothing against the government, and that it is good literature, or science, or poetry, as the case may be, and worthy of being published; it is then returned to the author, with permission to print it. The author's annoyance does not end here. He is obliged to leave a copy of his manuscript with the police, and a copy of the work as soon as printed, so that they may be compared, and any alterations or additions detected. If he desires to make any alterations in his manuscript after it has passed the censorship, he must send it back for a second examination. Persons holding responsible official situations are not exempt: if a censor himself wishes to publish anything, his manuscript must be given to the police that it may be examined by some other censor. All kinds of works, books of dry science not excepted, are subject to the censorship. To my great surprise, Endlicher, who gave me all this information, informed me that all the manuscript of his "Genera Plantarum" is sent to the police, who transmit it to Baron Jacquin, the censor for natural history, etc., and who is well paid for the business, but who knows just as much about it as if it were written in Arabic, and who certifies to each portion that it contains nothing hurtful to the people, nothing offensive to the emperor, to religion, etc., and more than all, that it is good science! To avoid the annoyance of sending it back repeatedly, as he has alterations to make, he is obliged to promise the printer to indemnify him, in case any discrepancy is observed between the manuscript and the printed work. Endlicher spoke of all this in terms which there is no necessity for me to record just at present. He gave me an anecdote respecting the publication of his earliest botanical work of any consequence, a Flora of his native town, the "Flora Posoniensis:" the manuscript being duly sent to Jacquin, that worthy refused to give it his imprimatur, because, it was arranged according to the natural system! which Jacquin did not like; and Endlicher was obliged to apply personally to the ministers and take great pains, when he obtained permission to print in spite of the censor; he took his revenge by dedicating the work to Baron Jacquin himself! This system sufficiently explains the low state of literature in Austria, as compared with northern Germany. I could hardly believe all I have heard, had I not obtained my information from such authentic sources....

Friday evening, 31st May, 1839.—The remainder of the morning was devoted to the botanical cabinet; and in the afternoon and early part of the evening I called with Endlicher upon Mr. Fenzl,[103] the aide-naturaliste in the botanical department, who is confined to his bed by some affection of one of his legs. He is engaged in a monograph of Alsineæ, which I think will be very faithfully done, and we looked over several collections by his bedside. I made a bundle of all I wished to examine, which are sent to my lodgings for the purpose, and which will give me occupation for the evening. He introduced me to his frau, a regular German lassie, and we managed to converse altogether for some time in a curious mixture of French, German, and English.

ON THE DANUBE, on board the Dampschiff
(steamboat) Maria-Anna, bound for Linz, 5th June.

Schönbrunn, the Versailles of Austria, is much like Versailles itself on a smaller scale, but much less magnificent. I visited the grounds with Endlicher, and also visited the botanic garden attached, under the care of M. Schott.[104] The garden is very finely arranged, but all that is particularly worth seeing is the conservatories and the large collection of exotics, many of them very old like those of Kew. It is richer than Kew in Palms, Aroideæ, etc., but in other things it seems not quite equal. As we passed by the palace, the emperor was pointed out to me, through the open windows of his cabinet. I am told privately that he is scarcely compos mentis, and that all government affairs are managed by a regency of which Metternich and Archduke Charles are chief. We went next to see Baron Hügel, and the extensive collection of living plants he has collected during his travels. I think I have not told you the cause of his long journeying. He was, it appears, the accepted lover of an accomplished and beautiful lady of very good family here, and their union was considered as a settled affair. But unfortunately for poor Hügel, Prince Metternich looked upon the lady and determined to have her. So he sent Hügel upon some humbugging political mission, to Paris I believe, and during his absence he made his propositions to the father and mother, who were not slow in discovering that Metternich, with all his riches and power, malgré his sixty-odd years, was the fittest bridegroom; and I am sorry to add that they persuaded the daughter to the same opinion, though she could have had little liking to the old fellow personally, and was said to be much attached to Hügel. The latter at length found out why he was sent to Paris, and came back with all speed, but he was too late. His intended became Princess Metternich, and Hügel set out to cure his disappointment or forget his love by traveling in foreign lands. Metternich, being glad to get rid of him, threw facilities in his way, and being fond of plants he collected and sent home an immense quantity for his garden. At the same time he made extensive collections of dried specimens, etc., which all reached Vienna safely. He spent nearly all his fortune in traveling, and would have been in a quandary, but the government, that is to say, Metternich, bought all his collections of dried plants, animals, etc., for the Imperial Cabinet, giving for them an immense price, some thirty times more than they are worth, and so Hügel is able to enlarge and embellish his place, improve his garden, and build most beautiful greenhouses. He has fitted up his house very tastefully, and filled it with all manner of strange things, arms, idols, and so forth. His collection of living plants is larger than that of Schönbrunn, though the trees are younger.

Several days after my arrival I called to pay my respects to our minister here, Mr. Muhlenberg, and the secretary of legation, Mr. Clay. Philip and myself also spent an evening at Mr. Clay's, where we met Mr. and Mrs. Muhlenberg, and their daughter, a young lady of about seventeen; also Mrs. Clay, a pretty woman, and Mr. Schwartz (the American consul here) and his wife, who both speak English indifferently well. Muhlenberg seems quite sick of living here, and speaks of the Austrians with anything but praise.

We went one evening to a public garden, of which there are many here, to hear the most celebrated musician here, Mr. Strauss. A few kreutzers are charged for admission, and the company are nearly all seated, at little tables, eating a substantial supper, or sipping coffee or ices, as they incline, while Strauss with his fine band played the finest music, mostly pieces of his own composition. It was the best music I ever heard.

Philip left me on Monday evening and went to Prague. On Tuesday I arranged passport, left parcels to be sent to Hamburg, took leave; came out to Nussdorf after dinner, from which the steamboat leaves, and after seeing my luggage deposited safely on board, I climbed the Leopoldsberg, a steep mountain between eight hundred and nine hundred feet high, and enjoyed the beautiful and extensive view from its summit,—a fine view of Vienna, of the Danube branching into many different streams, forming pretty green islands, and the whole of the broad valley far into Hungary. In a fine day, it is said the towers of Pressburg, forty miles off, may be distinguished. The Danube, which is here as large as the Niagara, broad and swift, washes the base of the mountains, and the view up the river, though not so extensive, is more picturesque. I collected a handful of plants, bid good-by to Vienna, and descended, slept on shore, and was on board the boat in time to start with it at five o'clock this morning.

This is the first time I have slept in a genuine German bed,—a feather-bed beneath, and an eider-down bed the only cover. It is inclosed in a sheet like a pillow-case, and under this you creep. In the winter it might do very well, but at this time of the year it is very oppressive. The upper sheet here I find, in all cases, is tied fast to the coverlet, which is all of one piece, and just long enough to cover a moderately sized man like myself from the chin to the toes. A taller person must choose between his shoulders and his toes, for they cannot both be covered.

Living is dear in Vienna. I stopped at a cheap hotel, being aware of this, and lived as economically as I well could, but I find I have made way with a very considerable sum. The only way to travel cheaply anywhere on the Continent is not to be in a hurry, and to understand the language.

Notabilia for Dr. T.—I have seen Corda[105] at Vienna. He is one of the curators of the collection at Prague, and was at Vienna on a visit. Learning that I was there, he called and left his card. I afterwards saw him at his hotel. He is a little fellow about thirty, with a small expressive countenance. He works chiefly at minute fungi, on which he is publishing a large work. I saw a part of it in London. He showed me an immense quantity of drawings, which he makes with great rapidity. He is also publishing a work supplementary to Sternberg's "Flora of the Former World," a work of which Corda did a good part. He gave me two copies of a lithograph of Count Sternberg,—now dead, as you know,—done by himself. I observe by his drawings that he has anticipated an unpublished discovery of Valentine's, which he showed to Lindley and myself in London, about the holes in the tissue of Sphagnum opening exteriorly. I looked at Corda's microscope (one of Shiek [?] at Berlin), but it is inferior to the English or Chevalier's.

I made a second visit to Fenzl, as he lay in bed; had a long botanical talk with him, and think him a most promising botanist.

Ungnadia (the character of which Endlicher has not yet published,—the last plate in the "Atakta") was named in memory of Baron Ungnade, once an ambassador from Austria to Constantinople or Persia, I forget which, and the first to introduce Æsculus Hippocastanum into Europe,—hence the propriety of the name. Endlicher is soon to publish the description in the "Annals of the Vienna Museum," which work, with the "Iconographia Generum Plantarum," he has promised to send to Hamburg for me, along with the parcels of plants given me. We have studied the new Loganiaccous plant from Florida. It proves, as Brown guessed, near his Logania § (or Gen.) Stomandra, but extremely distinct from that or any other genus, by the character of the style which Decaisne first noticed. Endlicher is to give a figure in "Iconographia Generum Plantarum," and the description has gone to the printer in one of Endlicher's articles in the "Annals of the Vienna Museum,"—Cœlostylis Loganioides, Torr. & Gr. Can't we get more of it? Has Leavenworth found it?

I have been looking over the "Reliquiæ Hænkeanæ," and examining what specimens of the collection from North America they have in the Vienna Herbarium. Endlicher goes this week to Carlsbad to recruit his health, stopping a day at Prague. He has kindly taken a list of my desiderata of the species published in that work, and I hope to get some bits of them. I have copied so much from the work that we can get along even if I do not see it again, but as I was about to purchase it, Endlicher suggested that he should see if Presl himself has not a copy left for us. Following this hint I have sent by Endlicher a copy of the "Flora" to Presl,[106] in nomine auctorum.

There is a new genus of Presl in Loaseæ (Acrolasia) from Mexico, which may be Nuttall's. The most curious thing is a new genus of Datisceae from Monterey (why have none of the other collectors found it?), called Tricerastes; very interesting.

I find from all inquiries that it is very difficult to find Nees von Esenbeck[107] at Breslau, especially in the summer. He is a queer stick altogether, is not well satisfied with his situation at Breslau, and spends the greater part of his time at a little place high up in the Riesengebirge, studying Hepaticæ.

I have bought Grisebach's new "Genera et Species Gentianearum," and have been studying it on my way in the steamboat. It seems very well done, particularly his preliminary matter on structure, affinities, development, geographical distribution, etc., which is very interesting. It is very carelessly printed. Our well-known "Tuckerton," in the pine-barrens, figures under the form of "Juckerten"! Let this suffice at present.

SALZBURG, June 10.

Arrived at Linz Friday noon, dined, looked a little about the town, which is remarkable for nothing except its agreeable situation on the Danube, and its unusual kind of fortification; and at half past one started for Gmünden, about thirty-five miles by railroad, in a car drawn by horses. This railroad, the oldest in Germany, is rather a primitive affair; we were jolted more than on the ordinary roads, which I have found everywhere excellent. The first part of the road was very uninteresting. I was seated in the middle of the car, with five or six inveterate German smokers around me, each equipped with a huge meerschaum pipe with a wooden stem nearly as long as your arm, which he replenished as often as it was exhausted, and all puffed away in concert as if they were locomotive engines and our progress depended upon their exertions. You are everywhere annoyed in the same way, but I have become accustomed to it so that it does not trouble me as at first. At length a fat military officer next me smoked himself to sleep; and I was amusing myself with the ridiculous pendulum-like motions he was making, his pipe still grasped by his mouth at one end and by his hand at the other, when he knocked his head against the window and pitched his hat into the road, to his great astonishment and our infinite amusement. We passed through Wels, and afterwards Lambach, a pretty place and most beautifully situated upon the Traun. In this part of the journey we had a fine view of the Salzburg Alps, which rise to their greatest height just where Austria proper and the provinces of Styria and Salzburg meet. From Lambach to the end of the journey, the country appeared completely American: finely wooded with fir and larch with here and there a clump of beech. We reached Gmünden just at twilight, a neat village on the very bank of the Gmündensee or Traunsee, for it is called by both names. The situation, close down upon the water and in the bosom of green undulating hills, is as lovely as can be conceived, and is in fine contrast with the upper extremity of the little lake, where the dark and lofty mountains rise abruptly from the very edge of the water, not leaving room enough even for a footpath. Their summits were still covered with patches of snow, but they are overtopped by the peaks of the Dachstein and other portions of these Alps which are crowned with perpetual snow. I found at the Goldenes Schiff neat rooms, and a most comfortable bed, which I was prepared fully to enjoy, having first made a supper on nice trout from the lake, with a few etceteras. At seven o'clock the next morning I was on board the little steamboat,—commanded by an Englishman, as most boats are in Austria,—which affords the only means of communication with the country beyond. The morning was pleasant, and I had a good opportunity of seeing the finest scenery I ever beheld; indeed I do not expect ever to see it surpassed. As we left the green slopes at Gmünden behind us, the mountains which inclose the upper portion of the lake gradually disclosed themselves more distinctly; halfway up, we were opposite the gigantic

Traunstein, whose naked and weather-beaten summit had been full in view almost ever since we left Linz the day before. It is a huge mountain, appearing as if split from top to bottom and turned with the cloven side toward the lake, so that it presents a perpendicular wall of jagged rock nearly three thousand feet high! leaving just room sufficient between it and the water for one or two fishermen's huts, which look the veriest pygmies. The mountains beyond this on the same side are equally picturesque, but not so high. They rise in sharp isolated peaks, leaving the wildest glens between, down which streams fed by the snows of the mountains in the background come leaping to the lake. On a promontory which seems from the lower part of the lake to form its southern extremity stands the little hamlet of Traunkirchen; the picturesque little church was founded by the Jesuits, who once had a small establishment here; a little nook is occupied with the wee bits of cabins belonging to the peasantry employed in the salt-works or in rowing the salt-barges down the lake; they are set down here and there, as room can be found, and add much to the beauty of the view. As the boat doubles this promontory, Gmünden and all the lower part of the lake is lost sight of, and you seem to be on another smaller but wilder lake, entirely shut in by the precipitous mountains; a few minutes more and we are landed at Ebensee, the little salt-village at the head, where the Traun enters, and you regret that the voyage is so short. I was strongly inclined to go back again with the boat, and return again in the afternoon; but knowing I had no time to lose, and that I might not readily find another convenient opportunity of going on to Ischl, I was obliged to bid farewell to Gmündensee. Loveliest, wildest of lakes, I shall not soon forget thee.

I had not time at Ebensee to look at the works where the brine is evaporated, which seem to be on a large scale. The brine is brought here in aqueducts, some fifteen or twenty-four miles, since fuel is more plenty here, and it is found more economical to bring the brine to the fuel than the fuel to the brine. The stellwagen was ready, and I took my seat. A ride of ten or eleven miles up the valley of the Traun, a narrow defile bordered by lofty mountains, brought us before noon to Ischl. It is a pretty village, lying in a green valley formed by the junction of the little river Ischl with the Traun; it contains extensive salt-works and is a favorite bathing-place, people of all degrees coming here in the summer to pickle themselves in the salt water. Three immense ridges of mountains come down almost into the village, leaving a triangular space for the village, with just three ways of getting in or out, viz., by ascending the river as we came, or by either the Ischl or the Traun as they enter the valley.

I took a hasty dinner, and left the hotel at one o'clock, determined to enjoy the satisfaction of climbing a real mountain. The Zeimitz, the highest in the neighborhood, is said to command the finest prospect, and it looked as if I could ascend it in an hour or two with the greatest ease, although the guide-book says that ten to twelve hours are necessary for going and returning. I have accomplished the task; I climbed the mountain, 5000 feet high, traveled over the snow from one to the other of its four peaks at considerable distance from each other; enjoyed the most magnificent prospect; filled my portfolio with alpine plants, descended the steepest side, picking my difficult way down the rocks and sliding down immense snowbanks, until I was past the alpine portion; then making a turn to a subalpine pasture, where cows and goats are driven to pass the summer, I struck an old path, and ran with all speed to the gorge at the base, where the stream that I had traced from its source as it trickled from a snowbank, and down a succession of little cataracts, was now a foaming and rushing torrent. It was then just twilight, and a quiet walk of an hour brought me back to the hotel at nine o'clock, quite proud of my feat and delighted with the fine view I had obtained. But I have paid well for it. In the morning I could scarcely stir for the aches and pains in my bones, and even now the extensor muscles of my legs are sore to the touch and bear woeful testimony to the hard service they have been obliged to perform. "I shall think about it," as Mr. Davis says, before I ascend another mountain.

And yet I feel myself well repaid for all my fatigue. To say nothing of the prospect opening out wider and grander as I ascended, I had from the summit a magnificent mountain panorama which it was well worth the labor to see; the summits of more than one peak white and brilliant with perpetual snow and ice. The most stupendous of all is the Thorstein or Dachstein, which closes the view to the south, with its immense glaciers of the most dazzling whiteness, from which numerous steep pinnacles rise like spires, towering high above all surrounding objects, illuminated by the rays of the setting sun long after all other objects are left in the shade. The dark lake of Hallstadt was distinctly seen, appearing to reach up to its very base. I could not distinguish the village which is hidden under the cliffs at that end of the lake, where from November to February the inhabitants do not see the sun, they are so shut in by high mountains. Four other lakes were in full view, two of them lying almost beneath my feet.

And then imagine my pleasure at collecting alpine plants for the first time, some of them in full blossom under the very edge of a snowbank. I filled my portfolio with Soldanella, Rhododendron, Primula Auricula, Ranunculus Thora, and another with white flowers, etc., etc. I am sorry to say that in my eagerness I have left my knife, last relic of the Expedition, and so long my trusty companion, somewhere on the top of the mountain. Sunday was at least a day of bodily rest, for I did not rise until past ten o'clock, and hobbled out but once beyond the limits of my hotel. I was obliged to leave, however, late in the evening, about half past ten, when the eilwagen, which comes but twice a week, arrived from Gratz on its way to Salzburg; and here I found myself at six o'clock this morning; a rainy day, and a very dull town, with nothing but its fortress and its exceedingly beautiful and romantic situation to make it interesting. There are many objects of great interest in the neighborhood, but this rainy day prevents any distant excursion; my place is taken for Munich for to-morrow morning, and not even the inducements of "the most beautiful region in all Germany," as it is called, not even the sublimities of the Berchtesgaden and the Königsee, which are but fifteen miles off, shall detain me longer. I begin to look with expectation toward the end of my journey, and have already in my plans shortened it a little. I have looked about the old churches and buildings of this town, and am waiting now for it to clear up that I may climb the Mönchsberg, and enjoy the prospect that is said to be so fine. At midday I had hopes of a pleasant afternoon, but it is now raining harder than ever.

In this region, as in the retired parts of Styria, through which I passed to Vienna, you are charmed with the kind-hearted simplicity of the people. If you meet them in walking, they always give you some word of greeting, and commonly take off their hats and bow to you; yet there seems to be nothing servile or cringing in it. You get a porter to carry your baggage, who, instead of asking for more when you have given him already more than he expected to receive, takes off his hat, makes you a low bow, and thanks you most heartily, though without any palaver. So with the servants, who never ask anything, and I suppose would not if you were to forget them altogether; I doubt if they would ever remind you; you give them about a third part of what an English servant would expect, and you have them all most heartily wishing you bon voyage or glückliche reise, according to the language they speak. In some places they say the chambermaid kisses your hand, but this has not happened to me yet. The women, when not rendered wholly masculine in appearance by performing the labor of men, which is very common, are almost universally good-looking, and in such vigorous health. I do not admire their head-dress, which is ordinarily a black silk thing tied closely around the head and tied in rather fantastic bows behind. The women of Linz and all this part of the Danube wear, when in full dress, a cap of tinsel or gold lace, shaped exactly like the Roman helmet, which fits close to the top of the head. But fashions never leave this world; when you ladies throw aside some mode, it is picked up and perpetuated in some out-of-the-way part of the world. Thus, for example, all the young fraus of Ischl wear balloon sleeves, after the most approved fashion some three or four years ago. I assure you it looked quite natural to see them again, even upon the buxom damsels of the Salzkammergut (there's a name for you).

It is now half past seven; and it is still raining most obstinately, so ascending the Mönchsberg is not to be thought of; and I must make up my mind to leave Salzburg without this view. My trunk is sent to the office of the brief-post-eilwagen, all ready for starting at six o'clock in the morning, and to-morrow evening at eleven I hope (D. V.) to be in Munich, seventy-eight miles. I owe Bentham a letter, and have not written him or any one else since I left Paris. I will take this convenient opportunity and write forthwith.

MUNICH, 12th June.

I arrived in this capital of Bavaria last evening at eleven o'clock, after a tedious, though not uninteresting ride of seventeen hours. The day proved a fine one, and after leaving Salzburg through the curious tunnel that penetrates the Mönchsberg we came abruptly into the open country; and as the mists gradually rose from the sides of the mountains and we ascended some small hills, I obtained some most beautiful and picturesque views of the surrounding mountains. The Stauffenberg, which stood between us and Berchtesgaden, a magnificent mountain, was for a long time the most prominent object; backed by the more distant central portions of the Salzburg Alps, all white with snow. It was only as I left this place that I could appreciate the beauty of its situation, and I felt a momentary regret that I had not stayed a day longer and visited Berchtesgaden. These fine mountains and those of the Tyrol (the more western portion of the same chain) were in full view during the whole journey, filling the southern horizon, while we journeyed through a rather level country; for the whole of Bavaria south of the Danube is a great plain, stretching from that river to these mountains that skirt its southern border. It is an inclined plain, since Munich, though in a perfectly flat region, is about sixteen hundred feet above the level of the sea. We crossed the frontier in an hour after we started, where our baggage was slightly and very civilly examined, and our passports viséd by the Bavarian police. We passed two pretty lakes, but no place of interest except Wasserburg, situated in a picturesque dell on the river Inn. For companions I had a Dane, who spoke a little English surprisingly well, and was very agreeable; a German, who spoke a little French; and a Frenchman, who had come up the Danube from Constantinople, and who tired us all with the continual clack of his very disagreeable voice. I took up my abode at the Schwarzer Adler, a very comfortable and quite cheap hotel; slept pretty well; rose early this morning to take a look at the town, which within these last twenty years has become a magnificent capital; saw many of the public buildings,—that is, their exterior,—churches, and squares; went to the office of the police and obtained the required permission de séjour; and then went to the Royal Cabinet to find Martius, for whom I had three letters of introduction. He is a small man, not so tall as I, quite thin, but rather good-looking, apparently fifty years old, but his hair may be prematurely gray. He seems to have his hands very full of business, but he received me with cordiality; took me to the library and the cabinet of natural history, which are in the same building, told me to amuse myself till one (the universal dinner hour), and meet him at the Botanic Garden at three, and afterwards spend the evening at his house. The cabinets here are in an old, rather inconvenient building, once a Jesuits' college, which now contains them all, as well as the library, the lecture-rooms of the university, etc., but in a year or so all will be removed to very fine buildings the king is erecting for their reception. Excepting the Brazilian collections, which are large and good, there is nothing worth particular notice in the zoölogical and mineralogical cabinets; they make no great show after that of Vienna. The library is immense, this and the one at Paris being the two largest in the world; the books fill a great number of rooms, none of them magnificent but very convenient; the whole is soon to be transferred to other quarters. I was introduced to one of the librarians, who was at the moment showing the curiosities of the collection, very old and rich manuscripts,—the earliest attempts at wood-engraving, etc.,—to a party of English. When he had done with them I told him he must have been bored quite sufficiently for once, and that I would not trouble him any further just then, but that I wished to acquire some useful information about the plan and arrangement of the library, rather than to see its curiosities. So he fixed upon Friday morning, when he would be quite disengaged, and would gladly afford me all the information I desired. Shortly after dinner I went down to the Botanic Garden; found Martius, who, having an unexpected engagement, consigned me to the head gardener, and I was very kindly shown over the whole establishment, which is much larger and better than I had supposed, and in excellent condition.

Afterwards I strolled about the town for an hour or two, heard the fine military band in the Hofgarten, and at half past six went to the house of Martius; saw his wife, who looks much younger than he, and I suspect he was not married until after his return from Brazil. She seems a very intelligent and pleasant lady, understands English pretty well, but does not speak it, while Martius speaks extremely well; the eldest daughter, a pretty girl of thirteen, speaks French fluently, has taken lessons in English, which she reads readily, but speaks slightly; there is another daughter of about ten, another still younger, and a boy a little more than a year old completes the list. Professor Zuccarini[108] was there, and afterwards an entomologist, whose name I forget, dropped in; also a young man from Rio Janeiro, a Dr. Hentz from Vienna, who inquired especially after Dr. Buck; the director of the music in the royal chapel here; and two ladies, one of whom sung exquisitely. The director and Dr. Hentz both played the piano to perfection, and, to crown all, Martius seized his fiddle, quite to my surprise, and played with great spirit. Before they were done a little crowd had began to assemble before the windows. So the evening passed off very pleasantly.

I like the sound of the German language much; it is manly, and certainly not more rough than the English. From the lips of the women and the little children I assure you it sounds very musical, and I often stop in the street to listen to it, when I do not understand a word that is spoken.

13th June, 1839.—I passed the whole morning, that is, until one o'clock, at the Botanical Cabinet, looking at grass and such like. After dinner Zuccarini called for me, took me to his house, showed me his Japan plants, the work he is publishing on them, etc. I looked over and named his American Cyperaceæ, and he made me most bountiful offers for exchange. He gave me some of his publications and even offered me his "Japan Flora" (Siebold's), which is an expensive work, but it is very desirable for us to have, though it will be rather difficult for me to give him an equivalent. It is now sunset, eight o'clock; all the shops in the town have been closed nearly an hour, the people all enjoying themselves in the gardens roundabout. I am going to bed early, in hopes to rise in time to go down to the Garden and hear Martius lecture at seven o'clock. He lectures every morning at that hour, and Zuccarini again every morning from eight to nine, and also from eleven to twelve. The scientific people here have been arranging a little fête for Saturday, the birthday of Linnæus. It is decided that there is to be a botanical excursion, I believe, to the Tegernsee, some fifteen miles off, and I suppose also a picnic dinner. I have not learned all the particulars, but this I shall do in time, as I am to be one of the party.

14th June, 1839.—I rose early this morning and went to hear Martius lecture at the Garden at seven o'clock. He is a good lecturer, fluent and clear. Called on Dr. Schultes;[109] then returned to breakfast; afterwards spent the morning at the cabinet, with the exception of an hour devoted to the library, which one of the chief officers very kindly showed me through. They have about half a million books, excluding duplicates, and about 16,000 manuscripts. The librarian took much pains to explain to me the arrangement and classification of the library, which is in excellent order, and to show me as many of the rarities as I desired to see: very ancient Greek and Latin manuscripts of the Bible or the Evangelists; a number of very old and richly illuminated German manuscripts; the collection of printed books without date, of which they had 6000 (these early printed books being many of them intended to pass for manuscripts); a copy of Faust's Bible again (the first book printed),—they have two; Luther's Bible, beautifully printed on vellum, and illuminated,—in the frontispiece his original portrait, a sturdy-looking old fellow, who looks as if he might have been as fearless as indeed he was; the portrait of Melanchthon, by the same artist, whose name I forget, is given on the next leaf. I saw also a manuscript letter of Luther, and many other things, too tedious to trouble you with now.

Dined with Martius and his very pleasant family; stayed until six o'clock, looking over plants, etc.; took a little walk, now that it is a little cooler, for the day has been exceedingly sultry, and am now going to bed, as I have to rise at half past four and meet the pedestrian portion of the Linnæan party at half past five. If it be as sultry a day as this has been we shall have warm work of it.

15th June, 1839.—We had a truly German fête champêtre, and I have learnt more of German life and manners in one day than I could otherwise have obtained in a long time. I was at the place of rendezvous at the time appointed, and met there the two professors and about thirty students, with whom we set out on our excursion, and our number was soon doubled by the accessions we received. Our course lay along the banks of the Isar (what lad that has been at school has not heard of "Isar rolling rapidly"), along which we ascended for about six miles, botanizing on the way. It was about twelve o'clock when we reached the place where the Linnæan celebrations are always held. Here we found Madame Martius and the girls, who had arrived in a carriage, and the lady and children of another professor. Three or four other professors also joined the party: Professor Tirsch, the celebrated Grecian scholar; Professor Neumann, of Oriental languages; a celebrated physician, and some others. We filled an immense rustic dinner-table spread in an open pavilion, ornamented in a simple manner with branches and flowers, and a portrait of Linnæus. Professor Martius then read his address, which I judged from its effects upon the audience to be humorous; then followed the dinner, plain but good, consisting of three or four courses, beer supplied ad libitum, and this was no trifle, as you would understand if you could see how all these Bavarians swill their beer. It is light, extremely light as compared with English. But you may judge how cheaply the Germans contrive to live, and how cheaply and simply they get up an affair which in England or at home would cost a round sum, when I inform you that the whole charge for dinner was twenty-four kreutzers or one Austrian zwanziger (sixteen cents!). This I suppose did not include the wine, of which there was a small supply, provided, perhaps, by Martius himself.

Three or four odes, written for the purpose, some in Latin, others in German, were sung, with a heartiness and a nicety of execution entirely German. Three or four toasts were drunk, some speeches made, and the party left the table. The greater part, excluding the ladies, then went to the Linnæan Oak, a young tree planted on the day of this fête five years ago. Here all took their seats on the grass around it, and a number of half-serious, half-humorous addresses or meditations were made, the people all sitting at their ease; then a song for the purpose was sung, and the celebration was over. Some part dispersed immediately, but the greater part assembled around our dinner-table, and heard some music from a paysanne, who accompanied her voice with an instrument like a guitar. Martius and Zuccarini had arranged to stay over night in the neighborhood to botanize to-morrow, and wished me to stay also, which I declined to do, but returned in a carriage with Madame Martius and the eldest daughter. We had a very agreeable ride and reached the city just as it grew dark. We had all day most beautiful views of the Bavarian Alps, which seemed close to us. The different professors spoke English with me, Professor Neumann, indeed, extremely well; were very polite to me, and I obtained much important information, and have put myself in the way to get still more. The whole affair was extremely well arranged. I have printed copies of a part of the odes, and a copy of the print of Linnæus, a very good lithograph, which was brought to the place and sold to the students for twenty-four kreutzers (sixteen cents) a copy. This is not the birthday of Linnæus; the 24th of May is the proper one, but it is not then pleasant in the country here.

18th June.—On Sunday I attended service in the Protestant church, a large and fine building, which was well filled. A part of the royal family are Protestants, but the king himself is a bigoted Catholic. The interior of the church is made to resemble a Catholic chapel as much as possible; the altar has a picture behind it, and a small crucifix stood upon the reading-desk. There was a very short liturgy, and singing in which all the congregation took part, as is always the case in Germany. The sermon which followed may have been very orthodox for all I know, for I could understand but a few words of it. I spent the remainder of the day in my own room....

Tuesday evening.—This morning I went to the cabinet of botany, to the library, and after dinner to Martius; looked over his Carices, etc. We then walked to the Garden, and afterward to the establishment for telescopes, etc., of the successors of Fraunhöfer, where I bought a very pretty little achromatic glass and a simple lens; looked at his workshop and collections, etc....

It is so long since I have seen your handwriting that I might forget it, but I met with it to-day very unexpectedly, you would never guess where! Even on labels of Carices in Martius' herbarium. After I get to Switzerland I shall count days until I see England again, from which there are but two steps home, on board a ship, and off again.

ZURICH, June 22, 1839.

In the afternoon I called on Dr. Schultes, who offered me a pretty little parcel of Egyptian plants. Made up my parcels and left them with

Martius, to be sent, with the things that he and Zuccarini are to add, to Hamburg, against my arrival there. Spent the evening at Martius' house, and took my leave of madame and Caroline. I gave Madame M. my copy of "Childe Harold," a very pretty one, which she seemed to value considerably. Martius I saw again the next morning at the cabinet, and took leave very affectionately; he kissing me tenderly, after the German fashion. Ask Dr. Torrey to look in the list and see if Martius is not an honorary member of the Lyceum, as I believe, but am not sure. If he is he knows it not. The Lyceum has also been remiss in sending him the "Annals," which should not be, as he has been a liberal contributor. His works give him much trouble since the death of the late king, who was his patron and subscribed toward the expense; the present king does nothing at all for Martius or for science anyway, so that poor Martius is a little embarrassed. Meanwhile he is pressed down with his duties as professor, director of the Botanic Garden, etc., for which he is most miserably paid.

The Botanic Garden is better arranged than any other I have seen on the Continent, except at Paris, and I have secured a copy of the plan. But I must break off with Munich.—Arrived at Lindan, on Lake of Constance, yesterday; a fine lake, but too large to show well; the shores only at the eastern end mountainous; the rest ordinary, and in high cultivation, dotted with thriving villages; took a steamboat after dinner for Constance....

ON THE RIGI, 25th June.

I must resume the thread of my narrative where I left it, at my entrance to Zurich. I did nothing that evening but look about the town, visit the old church where Zwingli, the earliest Swiss reformer, preached. The prettiest view is from the new stone bridge which is thrown across the Limmat just where it emerges from the lake. The stream, like all those that proceed from these lakes, is full, and clear almost as glass, of a fine blue tint; it rushes with great rapidity, but is still and even. The view extends up the lake to its middle, where a slight change in its direction intercepts further view; beyond rise some low mountains; a little farther a higher range overtops these, and these are again overlooked by the Alps of Glarns, Schwyz, etc., with thin tall peaks and brilliant glaciers. The shores of the lake are highly cultivated and thickly covered with little manufacturing villages. This is a Protestant canton. I attended church and heard a preacher who seemed to be very earnest, but as his language was an unknown tongue, there was little chance of my being edified, and I spent the remainder of the day at my room. The new hotel here is extremely good. Early yesterday morning I prepared myself for a pedestrian excursion over the finest mountain regions of Switzerland, which will take me about ten days, if I do not get tired of it and give it up. Not that I intend to walk all the way, which would be a great loss of time, but to avail myself of steamboats, etc., along lakes, and a diligence when I am on routes which they traverse, knowing full well that there will remain many weary and difficult miles that can only be passed by the pedestrian. So I have packed up my trunk and sent it on to Geneva, at the opposite corner of Switzerland. The garçon of the hotel purchased a knapsack for me.... Thus equipped, my knapsack on my back, the Guide to Switzerland in one pocket, and Keller's excellent map in the other, I set out on my travels in search of the sublime. At nine o'clock yesterday morning I left Zurich; took the steamboat down the lake as far as Horgen, some eight or ten miles, where I took a little lunch, and crossed the bridge into the little canton of Zug,—Catholic, as one soon finds out, by the crosses and beggars which abound by the wayside. Here the lofty Mont Pilate, with its sharp peaks, was in sight; it lies on the other side of Lake Lucerne. Soon after I saw the Lake of Zug, and soon after one o'clock I reached Zug, on the borders of the lake of the same name, the capital of the canton, a retired and lifeless village. I entered the best hotel well heated with my walk, which now amounted to about twelve miles. I obtained a plain but very good dinner of soup, the everlasting corned beef, fish, roast, and strawberries and cherries ad libitum; chatted French with the voluble kellnerinn (the demoiselle of the inn); paid my bill of two francs, and was again on my way. It was very warm, so I walked quite leisurely down the shore of the lake; the scenery growing every moment more picturesque, the Rigi rising at its foot on one side, bold and abrupt, the Rossberg on the other. (A sad tale belongs to this last, of which I had often read.) I reached Arth, the little village at the foot of the lake and of these two mountains, at half past four (seven miles); took more strawberries and milk, and at five o'clock commenced the ascent of the Rigi by the shortest but most difficult footpath. The landlord told me the ascent took four hours and a half. This, indeed, I accomplished, but found it a hard task. But the desire of witnessing the sunset from the top induced me to do my best. I had plenty of offers to relieve me of my knapsack, and at length, as I left the village, transferred it to the shoulders of a stout fellow, for it began to grow weighty. The poor fellow I think earned the ten batz he demanded (about thirty cents), though he did not seem to mind it much. The first third of the ascent the path is formed of steps like a staircase, and is very fatiguing. After we meet the road for mules or horses, which ascends from Goldau, it is not so difficult. Both in the ascent and from the summit, I had a full view of the vestiges of the awful landslip of the Rossberg; the vacant space of the mountain occupied by the portion that fell and the scarred surface of the path are most distinctly in view, and at the bottom of the valley lies the huge and unsightly and confused mass of rubbish which overwhelmed and buried the three villages of Goldau, Bussingen, and Rothen. This catastrophe took place in September, 1806. Several hundred houses and other buildings were destroyed; cattle in great number, and four hundred and fifty human beings perished....

But time is becoming precious, and I must tell you in a few words of the view from the summit of the Rigi, though description is wholly out of the question. The view from the Kulm, or peak, owes its great beauty and extent, not so much to the height of the mountain, which is only 5676 feet, as to its isolation, giving a clear view in every direction. It is also easy of access; ladies and persons who do not care to walk can ride up on horses or mules, by either side of the mountain. So there are great crowds here all the summer....

I was called in the morning at half past three to ascend the peak and watch the effect of sunrise upon the Alps and valleys. The morning proved quite favorable, though a little cloudy. The mountains, lakes, and valleys were all distinct, but looked cold. At length a blast from a wooden trumpet (a better instrument than you would think) announced sunrise, and the sun appeared between two strips of cloud, lighting up first the distant and high peaks and glaciers of the Bernese Alps, the Jungfrau, the Finster-Aarhorn, the Titlis, highest of all,—the white glaciers shining like burnished silver. Soon the serrated ridge of the gloomy Pilatus is lighted up; the dark valleys become more distinct; the lakes look brighter, and the broad valley toward the north stretches before you like a map, far as the eye can reach, covered with hamlets and villages, and diversified here and there with beautiful lakes....

Stanz, 25th June.... I intended to leave the Rigi by way of Wäggis on Lake Lucerne; to take there the steamboat as it passed at two o'clock, and go up the farther part of the lake, the Bay of Uri, and finding, if possible, the mail-courier at Fluellen, to go with him to the summit of the pass of St. Gotthard, return as far as Hospital, and cross by the pass of the Furca and the Grimsel to Grindelwald, etc. If you had Keller's fine map before you, it would be easy to trace this route, and to find out also where I now am. Without it you will not do it so easily. So having plenty of time, I stayed on the Rigi until noon, and then descended leisurely, having grown wise by experience, and knowing that the descent of a steep mountain is much worse for the legs and feet than the ascent. Besides, a little storm arose, and I took shelter under an overhanging rock, and amused myself in watching its progress down the lake, and in hearing the deep and prolonged echoes of the thunder as it was reverberated from peak to peak among the Alps. It was a scene to be remembered. And then the numerous ever-changing aspects of the mountains and lake as it cleared up! Saw the steamboat at a distance, and hastened to the foot of the mountain, when it soon became evident enough that the boat did not intend to touch there; so we took a boat and went out to meet it. But although we drew very near them as they passed, they did not choose to take the slightest notice of us, and I was obliged, in the middle of the lake, to consider what should be done in such a predicament. I had no intention of awaiting the return of the steamboat and going with her to Lucerne, thence to begin the route to-morrow; and for a few moments I was a little bothered. But fortunately a pedestrian like me is not at the mercy of steamboats and stagecoaches; and the high satisfaction one feels at his comparative independence is one of the great pleasures of this mode of locomotion, and goes far to compensate for the fatigue. I reflected that I might not find the courier at Fluellen, and in that case should have a prodigious journey, and moreover that I had clearly saved the money I should have paid. So, learning on hasty inquiry that a blind mountain path led from the opposite shore into the canton of Unterwalden to Stanz, etc.,—from whence I knew I could reach the Grimsel, and if I chose St. Gotthard, and that it was the nearest way to the Grindelwald and all the finest part of Switzerland,—I ordered the boat to take me to that shore, where I was accordingly left to shift for myself as well as I could. But then came on one of the ills that flesh is heir to, most especially in traveling,—I wanted my dinner! I stopped at a cottage, the only one in the vicinity, but found no one but a little girl, who stared at me as if she had never seen a civilized being; saw no chance of getting anything to eat, so I climbed the mountain, very steep, and almost without a path; it evidently had not been crossed before, this season. From the top I saw the bay and village of Buochs, and in the distance, Stanz, which I reached at six o'clock; found an inn which within was more comfortable than its exterior promised. I think I never enjoyed anything more than the piece of cold roast veal and coarse bread, and the plentiful dish of strawberries with excellent cream that followed. Now that I had got out of the ordinary route of travelers, I determined to visit the valley of Engelberg. I asked the landlord for a char-à-banc (as there is a good enough road for this vehicle) or a horse, to go this evening, but mine host seemed to have made up his mind that I should stay with him all night, and insisted that there would not be time for Engelberg. So not to disappoint him, I made up my mind to rest for the night, and sallied out to look at the village....

MEYRINGEN, 26th June.

I have accomplished a journey to-day, such as I think few pedestrians have ever surpassed, considering the difficulties of a great part of the way,—from Stanz to Engelberg, thirteen miles, then over a tremendous mountain, the Joch, 6890 feet high, among the snows and near the glaciers of the Titlis and the Wendenstock, and then by a long path, through the most sublime mountain gorge and valley of Engstlen, to Meyringen. The distance from Engelberg is reckoned at nine hours (they always reckon by hours here), which on ordinary routes would be thirty miles. I do not know how far it really is. I accomplished it between half past eleven A.M. and half past seven P.M., and am fatigued past all conception, completely done over, and my feet apparently spoiled. To-morrow, perhaps, I will tell you something about it.

GRINDELWALD, Thursday, half past five, 27th June.

I take the first leisure hour to resume my account. I find that I must have walked about thirty-four miles yesterday, making due allowance for the windings of the path. I commenced at five o'clock, reached Engelberg at nine, where I rested till half past eleven, and reached Meyringen, as I said before, at half past seven. The journey from Stanz is through a narrow but fertile valley inclosed by high and picturesque mountains for about seven miles, when the valley contracts, the mountains on each side rise to a great height into sharp and bare peaks, leaving barely room for the Aar to descend between. It forms, I may say, one continual cataract from Engelberg to this point. Before this pass is reached I had gone by some other mountains which were very remarkable; among them the Brisenstock, a ridge of rock like the upturned edge of a hatchet, some 6,000 feet high, and throwing up from one extremity a column of rock like a vast obelisk. The road, which is carried at considerable elevation along one side of this narrow valley, is not difficult, and exhibits the whole way the most sublime scenery. The Wallenstock rises on one side to the height of above 8,000 feet; and those on the other side are not less lofty. Presently the shining summit of the Titlis rises before you, surrounded by others scarcely less elevated. The Titlis is the highest of the Unterwalden Alps, 10,710 feet. You then arrive at a place where the Aar forms a series of cataracts in the bottom of the gorge, nearly a thousand feet below you; the opposite mountain exhibits an almost perpendicular wall of rock, nearly 6,000 feet high, and a little cataract formed by the melting snow above falls from the top to the bottom. Soon I entered the little valley of Engelberg, the most beautiful and picturesque I have seen, probably the finest in Switzerland; at least that of Meyringen and this of Grindelwald, where I am now writing, are not to be compared with it. I only wonder it is so little known. I think it not improbable that I am the first American that has visited it. It is far out of the ordinary routes, and though easily accessible with chars from Stanz, yet the three passes that lead out of it are excessively difficult footpaths. It is a green, sunshiny valley, having perhaps eighty acres of plain, but very rich pastures rise up the mountain-sides to some distance; it is entirely shut in by the high mountains that rise on every side; the Titlis rising abruptly on the south within a few yards of the village, and sending down its avalanches in the spring close to the houses. But the glaciers are so situated as to send their summer avalanches in the other direction, so that the hamlet is not in danger; the other mountains toward the south have the glaciers on their summits, but the peaks on the other sides present naked precipices. The Engelberg, from which the hamlet is named (angel-mountain) is a lofty mountain shaped like a slender cone, with the apex cut off obliquely. It rises almost within the valley, and presents a very curious appearance. The large convent stands just between the base of this mountain and the Titlis. Attached to it is a very large and fine church for such an out-of-the-world place. I stopped at the simple auberge of the Engel (angel); mine host could only speak or understand German and Italian, so that our communication took place mostly by signs and single words, I giving him the German names as far as I could of what I wished. I got a very comfortable lunch of cold roast meat; but I wanted some strawberries, and could not think of the German name, and had considerable difficulty. At length he seemed dubiously to comprehend what I wanted; he went out, and returned in a few moments with a fine dish of the article in question. Excellent cream is as common as need be; so I had a fine feast. I found that I was the first visitor here this season. I amused myself with looking over the travelers' book (which you always find) and reading the remarks of former visitors. An Englishman the summer before had ascended the highest peak of the Titlis. I afterwards saw that this could readily be done, as my route led me close to the top of the main body of the mountain.

To get into the valley of the Aar it was necessary to cross the Joch, a mountain connected with the Titlis, and almost as high. The pass between the two mountains is almost 7,000 feet at the summit, is covered with snow, and is in immediate proximity with the glaciers of the Titlis. The ascent is exceedingly difficult; indeed, from all I can learn, it is much more difficult than any of the passes at all frequented by travelers. I took a guide to the summit and some distance beyond, as a stranger could never have found the way. My guide was an old man of sixty years. From a high ridge near the summit, which belonged rather to the Titlis, I had a magnificent view of the mountains to the north and the valley I had passed through, and on the other side, close to us, of a vast glacier; the streams emerging from it formed a small river, which we had some difficulty in crossing, and which emptied into a dark alpine lake just below. Here I gathered a few alpine plants, as souvenirs of the place. Another weary climb over the snow brought us to the top of the Joch, and here, where shelter was impossible, we were exposed to a shower, but our umbrellas protected us in part, and the view repaid for a little wetting. Descending a little, my guide showed me a lake almost surrounded with snow, fed by the glaciers; the outlet, the source of one branch of the Aar, was the stream which flowed down the valley I was to descend to Meyringen; the knapsack was again transferred to my shoulders and I was left to myself. As I entered the valley of Engstlen the scenery grew wonderfully fine. Tired as I was I enjoyed the whole journey extremely, though it took me four hours and a half of continual descent; yet I look back upon it with delight. The main stream formed a succession of beautiful cascades; the mountains on each side very high, and mostly perpendicular faces of rock, and down these a great multitude of cascades of all sizes fell, some of them springing 500 feet at a leap; others, falling from much greater height over the rocks, looked like long skeins of yarn, if you will pardon the simile, dangling in the air. It must be much like the valley of the Lauterbrunnen, according to the description; but I think the latter cannot excel it. I hope to know to-morrow. A shower drove me into a miserable châlet, the highest one inhabited at this season, where I found a young man, who dwelt there for the summer, with his herd of goats, and his brother, a young lad of fifteen, who had come up from Meyringen to bring him some food, etc., and was just about to return, I drank about a quart of milk fresh from the goat, and found it excellent. When it stopped raining the youngster and I started together; I transferred my knapsack to his shoulders, and a franc and a half to his pocket, to the great satisfaction of both parties. He proved a very useful little fellow, though I could not understand much of what he said; he showed me some waterfalls and curious things that I should otherwise have missed. With the true spirit of his nation, ever ready to improve an opportunity, he told me he had a brother who spoke French, who would be my guide for the next day. It rained most of the way, but I was compensated for the partial wetting by the views of the most beautiful waterfalls, which fell into the valley in great profusion from the high precipices on each side. I could sometimes see twenty at one view. After a long and weary descent we came at last near the bottom, where this valley, and two others almost at the same point, fell into the main valley of the Aar, and I could look at the same moment up four deep and wild mountain valleys. Then skirting along the side of the mountain, we soon descended to Meyringen, deep in the main valley of the Aar, with two fine cascades behind it, and another very fine one, the cascade of the Reichenbach, on the opposite side of the valley. Glad enough was I when we reached the door of the humble auberge, and great was the havoc I made with the eatables which the kind landlady provided in abundance and of excellent quality. I sat down on a sofa in my chamber to read a little, but fell asleep instantly; slept until eleven, then took my bed and slept until half past seven in the morning.

I can say, with Sancho Panza, "Blest be the man who first invented sleep." In the evening, what with my great fatigue and blistered feet, I supposed I should be scarcely able to move the next day, and that traveling on foot would be impossible. But I awoke perfectly restored, my limbs supple and my feet much better than I had anticipated; my guide made his appearance while I was at breakfast; said that it would take three days to make the excursion over the Great Scheideck to Grindelwald, then over the Lesser to the Wengern Alp, to Lauterbrunnen, and back to Meyringen by

Interlaken and the Lake of Brienz. I insisted that it should be done in two, with the aid of a char from Brienz, at the end of the second day. Leaving my knapsack here, and taking a few things in our pockets, we set out at half past nine; stopped on our way to see the falls of the Reichenbach, where the stream of the valley we were climbing makes the descent of 2,000 feet in a succession of leaps; the longest forms the celebrated falls,—very fine. Farther above numerous waterfalls are seen dangling from the perpendicular sides of the narrow valley; one, remarkably high and slender, is called the Seilbach (rope-fall). Ascended through beautiful mountain pastures, dotted with châlets; the peak of the Wetterhorn in full view directly before us, a sharp pyramid, one side dark rock, the other pure white snow. The body of the mountain was still hidden by the Wellhorn, the first of the chain of high Bernese Alps we were approaching (9,500 feet); then the Engelhörner (angel's-peaks) and high up between these, we had a fine distant view of the most beautiful glacier in Switzerland, the Rosenlaui, celebrated above all others for the purity of its untarnished white surface, and the clear azure of its depths and caverns. Stopped at a little inn, which is occupied only through the summer; got an excellent little dinner at half past eleven, charges moderate; visited another waterfall, and then walked half an hour out of our way to the foot of the Rosenlaui glacier, which descends to only 4,200 feet above the level of the sea; found a party there, two gentlemen and lady, the latter carried in a chair; admired the pure white surface, entered a little way into one of the crevices, looked down into the deep azure chasms; returning, viewed the awful gorge through which the stream from the glaciers makes its way, at least 500 feet deep, and only four or five feet wide, the water rushing and boiling and roaring in the bottom like mad. Threw down a big stone, and heard it crashing against the sides and shattered to atoms. Continued up the Scheideck, close along the broad and vast perpendicular side of the Wetterhorn; finally reached the summit of the pass (6,040 feet), and enjoyed the magnificent view of the mountains down the valley of the Grindelwald. The Wetterhorn (peak of tempests) rises, one vast precipice of alpine limestone, its base extending from Grindelwald on the one side almost to Rosenlaui on the other, and so near us that it seemed easy for a strong man to throw a stone against it, though it is really more than a mile off; its summit is 11,450 feet above the sea; this precipice consequently forms a wall about 6,000 feet in height. Next to this is the Mettenberg (perhaps 10,000 feet); and next, the great Eiger (giant, 12,220 feet), presenting its long thin edge, like the blade of a hatchet turned up into the air; while back of the Mettenberg appears the pointed cone of the Schreckhorn (the peak of terror, 12,500 feet). The vast space between these peaks is filled by an immense glacier, here and there interrupted, which under various names extends from Rosenlaui and Grindelwald almost to the Grimsel, and to Brieg in the Valais. The increasing supply of ice and the refrigeration of such an immense quantity forces branches down the valleys far below the level of perpetual snow, particularly these at Grindelwald, the lowest known; the base of the lowermost being little more than 3,000 feet above sea-level. I descended rapidly, looked down upon the two glaciers just mentioned, reached the little hamlet of Grindelwald in the bottom of the valley, close at the feet of these vast mountains, and a little above the foot of the lower glacier, which is so close that it seems almost possible to throw a stone to it; but I believe it is a mile off; reached here at five o'clock (twenty-one miles), having walked very deliberately. It is now just at sunset; the day has been warm; but now it is very cold, and I am shivering too much to hold my pen; besides, it is time for supper, and I want another view of the mountains. Adieu....

Villeneuve, 4th July, 1839.... Being unexpectedly detained here for a few hours, almost at the close of my Swiss pilgrimage, I resume my pen, which I have had no time to use for some time past, and must bring up my journal in a hurried way to the present. Since I broke off I have seen more than half the wonders of Switzerland. I can only now tell you where I have been from day to day; but I shall have much to give you viva voce some of the evenings of the rapidly approaching autumn. Stayed at Grindelwald Thursday night (a week ago); watched the clouds striking against the Wetterhorn and the Eiger and rolling down its sides; terribly cold. Friday, 28th, rose at four; started at five, in fine walking trim, after paying an exorbitant bill for very indifferent fare; was very confident that the guide paid nothing, and therefore suspected a connivance between him and the aubergiste to put all on my shoulders,—one of the evils of a guide; they are worse than useless on all the usual routes, indeed anywhere, except in ascending very high mountains and crossing glaciers; felt a little inclined to punish my guide, and therefore set off at a swinging pace and took him up the Little Scheideck much more rapidly than he ever went before. I buttoned up my coat and pretended not to be making any effort at all, while the poor fellow stripped off first his coat, then his waistcoat, the perspiration running off his face; until finally he pronounced it impossible to keep near me, and lagged far behind. At length I took pity on him and walked slower, but we crossed the Scheideck and reached the Wengern Alp, a journey of four hours and a half, in a little less than three....

From the crest of the Little Scheideck (6,300 feet) I got my first near view of the remainder of the high Bernese Alps,—the Mönch (12,660 feet), the Jungfrau (12,670 feet) (I have been giving you the height all along in French feet, as they are put down in Keller; in English feet the numbers will be considerably higher), with the two white peaks, the Silberhörner (silver-peaks), which belong to it.

Still beyond, though not quite so lofty, were the Grosshorn, the Breithorn, etc. The point where I stood commanded nearly the whole view, from the Engelhörner, Wetterhorn, a glimpse of the Schreckhorn, the Mettenberg, Eiger, Mönch, and Jungfrau, as I stood just in the mid-distance; an unsurpassed view it is. As I descended the other side to the Wengern Alp I lost those more to the east, but came still nearer to the Jungfrau....

At the Jungfrau hotel, a mere châlet on the side of the Wengern Alp, we were close under that magnificent mountain, separated only by a narrow gorge, and elevated just enough to have the most perfect view from base to summit. We had heard the day previous the crash and roar of falling avalanches on the other side of the Wetterhorn, and I was very anxious to see one; before long I saw two, one of them a pretty good one, come tumbling and roaring down the Jungfrau. Soon a thick cloud came and enveloped these mountains, so that I departed earlier than I should have done; it threatened to rain; and we descended into the valley of Lauterbrunnen, which is very deep and narrow, and had on the way a fine view of the valley and the mountains and glaciers that close its upper extremity. Saw the celebrated fall of the Staubbach, and was disappointed in it....

Walked rapidly down the valley of Lauterbrunnen to the lake of Brienz, turning aside so as not to pass through Interlaken, which is a little British colony; took a boat to the opposite end of the lake (eight miles); had a heavy shower and much wind; saw the falls of Giessbach from the lake, seven very fine cascades one above the other. Landed at Brienz; took a char up to Meyringen again, looking at the beautiful waterfalls from each side of the valley, now very full from the rains. Arrived at my own lodgings at five o'clock, having accomplished in the twelve hours fifty miles, of which thirty-two were traveled on foot.

Saturday, 29th, rose in good condition, breakfasted, and parted with my thoroughly Swiss landlady at five o'clock; went up the vale of Hassli, one of the finest in Switzerland, for the Grimsel, perhaps the wildest and grandest pass across the Alps. It is a footpath, or at best a bridle-path. I set out alone, with my knapsack on my back. Ascended a considerable distance when the clouds sunk lower and it began to rain, though I had the satisfaction to see down the valley that the sun was shining at Meyringen. Passed the last little village (Guttannen), a lonely place; above, the scenery grew to the very height of gloomy grandeur: immense blackened granitic mountains, clothed at the base with black stunted firs, above all naked tremendous rocks and peaks; between, just room enough for the river to tumble along, forming here and there a cataract. The view was heightened much, I doubt not, by the clouds and storm, so entirely in character with the scenery. I never before enjoyed a lonely rainy walk so much.

At the height of about 4,500 feet, and in the midst of the very wildest and most lonely scenery, reached the falls of the Aar at Handek, the finest in Switzerland,—indeed the only sublime waterfall here; viewed it first from below, then from the rude bridge thrown across just a few feet above where it leaps into the awful gorge. The scenery and all is in character, and for savage grandeur I have seen nothing to compare with it. Stopped at the châlet near the only dwelling within some miles; waited a little for the rain to subside, and finding that even here a traveler's first wants had been pretty well provided for, I made an early but most excellent dinner upon bread, butter, cheese, and honey, the last especially excellent. No signs of better weather; so started on, passing a spot where falling avalanches every winter and spring had swept over a vast space of rock and completely worn it smooth; was now above trees, with here and there a bit of scanty vegetation, but almost every step to the end was now on rock or snow, and I walked on to the hospice near the summit in the midst of a snowstorm, one and a half hours; knowing it could scarcely accumulate sufficiently to obstruct or obscure entirely the path until I could reach the place of shelter, I enjoyed it intensely, but had quite enough when, at one o'clock, I reached the hospice (twenty miles), near the summit of the pass, surrounded with unmelted snow, more than 6,000 English feet above the sea. It is as comfortable a place as can be expected in such a situation, now kept as a kind of inn during the summer, and in winter left in charge of a single servant, with a store of provisions to last him until spring. The winter before last it was crushed by an avalanche, but the man and his dog escaped, and reached Meyringen in safety. It is now repaired; the stone walls are extremely thick, the roof protected against the winds, as is usual here, by laying huge stones upon it. Laid aside part of my wet clothes, and lay down before the fire to dry the remainder; fell asleep; on waking had just begun to write, but when I had given the heading, in came three more travelers: two Germans, whom I had met before at Grindelwald, and a young Englishman; all thoroughly wet with the storm, which was now more violent. We all had to huddle about the fire, so there was an end of writing.

Awoke Sunday morning and found myself in mid-winter; very cold, snowing hard, and the wind howling frightfully around our humble but snug place of refuge. The other travelers determined to prosecute their journey, spite of the Sabbath or the storm, and to go by way of the glacier of the Rhone, the other side of the summit of the pass and about four miles distant. They sallied out with their guide and left me to myself, which was one advantage. But in three hours they returned, giving an alarming account of the difficulties and dangers of the way. When just abandoning the attempt they heard a cry for help, and succeeded in rescuing another party of three with their guide, who had lost their way in the thick mist and storm and were wandering about in the drifts, suffering extremely with the cold, and who, as well as their guide, had given up all hope of reaching the hospice unless their cries should perchance be heard and bring them aid. All returned to the hospice together, and no further attempts to leave it were made that day. When left alone I had the fire to myself, and was spending the time in as profitable a manner as possible, thinking a little, too, of the strangeness of passing the day in such an elevated position; so their return, with an accession to their company, though very desirable for them, was not so favorable to me. And then of all people in the world the Germans are the noisiest talkers; Frenchmen are nothing to them; the fire which dried their clothes and warmed their fingers loosened their tongues, and they kept up a continual gabble for the greater part of the day. Scarcely a winter passes that some persons are not lost in this pass during such storms. A gloomy lake on the summit of the mountain, into which the bodies are thrown for burial, receives the name of "The Lake of the Dead" (Todten-See).

Monday morning, still enveloped in the clouds, but the storm apparently over. Found it no use trying to make a visit to the Rhone glacier; the clouds were so thick we could scarcely hope to find it, and the recent snow so deep nothing could be seen. Was disappointed also by these same clouds in getting a view of the high Bernese Alps, particularly Finster-Aarhorn and the glaciers, from this side, but determined not to wait here longer; so set off at half past ten in company with a native of Valais, who was traveling towards home and served as guide; traveled through deep snow, climbed up to the summit of the pass, more than a thousand feet higher, where at first we were so completely enveloped in the clouds that we seemed actually to be traveling through them and on them; dug a specimen or two of Soldanella out of the snow to serve as souvenirs. At length the wind arose and now and then sent a hole in the clouds to give me some glimpses of the desolate yet grand scenery through which we were passing. Soon I got a view of the valley of the Rhone almost at its commencement, with the river flowing through like a mere rivulet; looked down upon Oberwald, the highest village in Valais, a collection of little châlets all huddled together as if to keep themselves warm,—as indeed they have need; got out of winter and snow and into the valley at the little village of Obergesteln, and walked, on the same day, through a quick succession of most retired little Swiss villages of the humblest sort, to Brieg, on the Simplon road, near the mountain of that name, which I reached at nine o'clock in the evening, making a journey of forty miles, a portion through the snow, in ten hours and a half. I would like to tell you much about the upper Valais, a region seldom visited by travelers, but have not time; people kind and simple; got nothing to eat on the way except hard and dry brown bread, that may have been baked ten days; passed the villages where avalanches had fallen in former years and crushed many people; the scenery much more picturesque than I expected, but was most interested in the people and their little villages; women mowing, reaping, and doing every sort of the hardest labor; all awfully afflicted with goitre, scarce a person wholly free from it; actually saw one woman with a goitre not quite as large as her own head certainly, but about the size of that of the child she held in her arms, apparently a year old; saw one cretin. Stopped a few moments at the principal auberge in the village of Viesch; found the priest with two of his parishioners playing a game of cards together. A stranger being a curiosity in that region, one person accosted me very politely, and took me up the valley a little way to see the glacier and mountains. Reached Brieg utterly worn out, but got a good supper and bed; this being just where the famous Simplon road commences the ascent of the mountains, there are many travelers and a good hotel, though dear.

Rose Tuesday morning at four o'clock; my feet and legs very stiff and sore; thought of going up the Simplon road into the mountains to see some of the galleries and bridges and get fine views, but the morning was cloudy and I did not like to lose the time; started off down the valley, but got on slowly and very painfully; however, walked as far as Lenk, I believe about twenty-four miles, and there hired a char, which took me on to Siou, the capital of the canton, about twenty-two miles further, where I slept.

Wednesday, rose at four, and feeling pretty stout, I started off at five on foot, and though certainly in very far from the best condition for walking, went on to Martigny to breakfast, which place I reached at half past ten, twenty-four miles according to the guide-book, but the latter part was very painful. From this place one may go to the Hospice of St. Bernard in ten hours. I would have been glad to have seen so famous a place, but as to scenery it is decidedly inferior to much I had already seen. One may go to Chamouni in nine hours, getting the superb view of Mont Blanc from the summit of Col de Balme on the way. Thinking it impossible to walk farther, I hired a mule, and a person with him, and went up to the top of Col de Balme (five hours), passing the vale and glacier of Trient. Reached the summit at four o'clock; enjoyed a fine view of Mont Blanc and its attendant peaks from top to bottom, or rather at top and bottom, for there was a belt of cloud about the middle,—a most superb and complete view, Mer de Glace and all.

Quite satisfied without going to Chamouni, so returned to Martigny at eight P.M.; another good day's work, particularly as I walked both up and down the worst part of the road, being merciful to the beast. On my descent obtained a splendid view of the Bernese Alps. Much aroused at looking over the register at the hotel, where the travelers expressed their opinions of the different hotels on the road, praising some, and speaking of others in terms of great reprobation; good plan. I think if the proprietor of the hotel at Sion (a very dirty hotel) could read all that is written in his own book he would burn it.... Lay down and slept till midnight.

Thursday, took diligence at one o'clock A.M. for Villeneuve; saw the falls of the Sallanches by moonlight; arrived at Villeneuve at half past seven, just after the morning steamboat had left for Geneva; am confident we were delayed on purpose, to induce us to go on in the diligence instead of the next boat. For myself I did not mind waiting till one o'clock, that I might make myself look a little decent, though I had not the means here of improving my appearance much; as to my boots, and indeed all my habiliments, they were much in the condition of those of the Gibeonites when they made their visit to Joshua. Wrote a little, went out to take a look at the Castle of Chillon, which is near,—the building itself not remarkable, but the situation fine....

Took the steamboat in the afternoon; passed Vevay, Lausanne, etc., etc., and after traversing the whole length of this much-admired, most beautiful lake, arrived at Geneva, just at sunset; having accomplished my pedestrian tour (long to be remembered) in ten days (excluding the Sunday)....

GENEVA, 19th July.

My mornings, between eleven and four, have been constantly and fully occupied at De Candolle's. Earlier in the morning I have spent much time with Mr. Duby,[110] a botanist and clergyman,—one of the government pastors here, and it is said almost the only one who is a pious man. I have yet to pack up a box of my gatherings and to send to the roulage to be forwarded to New York. I have taken lodgings, for my short stay here, with the Wolff family, very pious and excellent people, who are pretty well known to many persons of the same class in New York. One of the daughters is the wife of Dr. Buck,[111] and I believe your dear mother is acquainted with her. After dinner I have sometimes made little excursions in the neighborhood; once or twice I have been accompanied by Madame Wolff and the two daughters. They are very fond of walking, and often make long excursions on foot. The two daughters walk as fast as I can, and in fact one of them nearly tired me down the other day, when we were hurrying in order to watch the effect of the setting sun on Mont Blanc. I have taken quite a fancy to this river, the Rhone. I made my acquaintance with it when it was but a babbling brook; I have trudged along with it for many a mile, until it grew to a headstrong stream, and became so turbulent and muddy that it was obliged to jump into the lake to wash itself clean, and when it leaves the lake it is as clear as crystal,—emerald, I should say, for it is about that color. A few months ago I saw the same river in its old age, just falling into the ocean. Walked back along the shore of the lake; reached the house just in time to join in the evening worship,—a sweet hymn was sung (in French), one of the young ladies leading with the piano and all joining with their voices, and hearts, too, I doubt not; and then the venerable old man read a chapter, which I could understand very well, and closed with a simple and fervent prayer. You cannot know yourself how pleasant it is, after being jolted about in the rude world for months, to get again with a pious family. The house is just without the town, surrounded with a large garden and fine trees and shrubbery, and all very pleasant. Some days after, we made another excursion to visit their pastor. He was not at home, so I missed him, but saw his pretty garden. On the two Sundays I have heard one of the pastors of the Evangelical Society preach in the morning, and the clergyman of the English chapel in the afternoon. I have also had the satisfaction of seeing Mr. Malan, who, when he called here the other day, was so good as to hold a long and edifying religions conversation with me. He is a very apostle in appearance, and in conversation. Indeed, I have been thrown here into the midst of religious society of a high tone and of great sweetness and simplicity. I hope I have received some benefit from it. As I leave here I shall lose all this and shall see nothing more like it until I get home again....

TO GEORGE P. PUTNAM.

BÂLE, July 23d.

... I left on Saturday morning for Lausanne and Freiburg, where I heard the big organ on Sunday; came on in the night to Berne, and yesterday to this place over the Jura. I wished here to see Professor Meisner, but found out this morning, some hours after the steamboat had left, that he was absent on a journey. I was a great fool for not finding that out last night, in which case I should now have been below Strasburg,—and this evening at Mannheim. As it is, I can't wait here till Thursday morning for the next boat, and shall leave this evening for Schaffhausen and Tübingen, and thence push on, the best way I can, for Dresden and Leipsic. I do not lose a moment of time. Do not be surprised if I drop in upon you about the 4th or 5th of September. I would like to sail for home the latter part of that month. In early winter we will hope to give you an entire volume of "Flora," and see what you can do with it. I have blocked out, in my mind, scientific labor enough for several years to come, and several works some of which will be good in a publisher's acceptance of the term; others, I dare say, not. As Murray's fame is derived from Byron, so shall you be immortalized and known to all posterity as the publisher of the celebrated Dr. Gray!!!

We have not much time to lose, and on my arrival at London I shall be wonderfully busy. I hope you will have picked up a great quantity of books for me by that time. My future credit and comfort will very much depend upon my bringing home an immense quantity of books for my money.... When I was in England I could scarcely hold up my head as a Yankee should—what with our border wars and domestic quarrels. But now I feel greatly relieved. The recent "Birmingham affair" and several other things fortunately (?) give me "wherewith to answer them that are of the contrary part." Let them shut their mouths now! You know my address at Berlin, or you may address poste restante if you will. I think I shall be there till about the 25th August. I shall stop a few days at Hamburg. I think I may say that I shall not go up to Rostock. You will perhaps be receiving some letters for me, which, now you know my movements, you will act according to discretion either in forwarding to me or in retaining.

I have bought scarcely any books since I left Paris. I have had some good ones given me.

Excuse this hurried epistle. I have precious little time, and I find I am growing more and more slovenly every day. Adieu.

Most truly yours,
A. GRAY.

TO GEORGE BENTHAM.

... Arrived at Geneva by way of Villeneuve and the Lake. De Candolle and Alphonse had returned only three days previous to my arrival. They received me very cordially, and I went through the herbarium as far as the "Prodromus" is prepared.

From Geneva I went to Lausanne and Freiburg; ... thence to Berne, where I made no stay; thence to Bâle, to Schaffhausen, to Tübingen, where I spent the morning with Mohl;[112] reached Stuttgart toward evening and Heidelberg the next morning. Frankfort in the evening; took the eilwagen the same night for Leipsic; saw Pöppig,[113] Schwägrichen,[114] etc.; railroad to Dresden; saw Reichenbach[115] for a few moments, as he went into the country the same day; visited the picture-gallery, which deserves to be called the richest out of Italy; returned to Leipsic; to Halle; passed a day or two with Schlechtendal;[116] saw the Carices in the herbarium of Schkuhr;[117] Potsdam, Sans-Souci, the marble palace, the beautiful statue of the late queen of Prussia by Rauch (the second and best one); and thence to Berlin, where I remained nearly a month; saw the botanists, etc.

TO WILLIAM J. HOOKER.

LONDON, September 13.

MY DEAR FRIEND,—The "penny postage system" not being yet in operation, I embrace an opportunity that offers to send you a line in Pamphlin's parcels. I am again in London, you see; indeed I have been here about a week. But it is only to-day that I have had intelligence of your return to Scotland. I had some hopes that I should find you in London on my arrival, or that you would return here from Chatham, and that I should have the gratification of seeing you once more. I received your welcome letter of August 14th, at Berlin, for which I thank you much. I wish my friends at home were half as prompt correspondents. While on the Continent I have received precious few letters.

I have been much interested at Berlin, and worked hard. The herbarium of Willdenow is larger and in better condition than I supposed, and the general herbarium is very interesting and rich. Klotzsch[118] is very industrious, and has got the whole collection in much better order than most of the herbaria on the Continent. I am under great obligations to Dr. Klotzsch, who not only afforded me every facility at the Herbarium, but most cheerfully aided me in every possible way, and during a transient illness (for I was confined to my room for a week or so, and to my bed for a few days) he procured for me the best medical advice, and took a great deal of trouble on my account.

I lost some time by this, but fortunately I had nearly finished my work at the Herbarium, and afterwards I had a few days to finish, and to look at Kunth's[119] herbarium, with which I was rather disappointed. Kunth was extremely polite and attentive to me. He is at work upon the third volume of his "Enumeratio," but I fear it will not be very well done. I saw Ehrenberg[120] frequently, and Link[121] once or twice, but nearly all my time was spent at Schönberg, where the Botanic Garden and Herbarium are situated, which is nearly a half hour's ride from the city. The garden is much the finest in Germany, and the government annually expends very large sums upon it. The building exclusively devoted to the herbarium is very commodious, though Klotzsch begins to complain that he has not sufficient room. It is so far from town that there are no loungers there, and one may study perfectly undisturbed. I brought a few things for you from Klotzsch and Link, which Pamphlin is to send to-morrow.

Having lost some time by illness I did not go to Rostock, a most out-of-the-world place, although I suppose I shall hereafter regret that I did not see Lamarck's herbarium.

I spent several days at Hamburg, saw Lehmann, his herbarium, and the botanic garden; and took steamboat for London. Since my return I have been busily occupied in the city, completing some purchases for the

Michigan University, and shall be mostly thus employed during the remainder of my stay....

19th September.—I saw Dr. Richardson the day before yesterday, who informed me that the Erebus was still lying at Chatham, and (what I was not aware of) that I could reach Chatham in three or four hours. So I arranged at once to go down and see Joseph before he started, but the next day I learned that the vessels had dropped down from that port.

I expect to sail in the Toronto from Portsmouth on the 1st October.... I have yet very much to do. Yesterday I dined with Dr. Lindley and visited the Garden. One wing of the conservatory is erected and nearly covered with glass. It is entirely glass and iron, about 130 feet long, and will be very fine.... Believe me, my very dear friend, most truly yours,

A. GRAY.

NEW YORK, 5th November, 1839.

MY DEAR FATHER,—Through the favors of a kind Providence, my journey is safely brought to a close. I am happy to inform you that I reached New York last evening in the ship Toronto, after a passage of thirty-five days. I left London on the last of September, and Portsmouth on the 1st ult. The steamship Great Western, which left on the 19th of last month, reached New York two days before us! Our voyage was a rather pleasant one, although we had nearly forty passengers. It was rather rough, but no very hard gales. I was sea-sick but a single day, and then but slightly. I have brought with me nearly the full amount of my purchases of books for the Michigan library, a large collection. I am waiting to hear from Detroit to know whether it will be necessary for me to go up there this fall. I hope I shall not be obliged to make this journey until spring. I shall not come up to see you until I hear from Michigan, when I can take Sauquoit in my way if it be necessary to go to Michigan. I am now busy in getting my boxes and parcels through the custom-house, which is a tedious business. I hope I shall be allowed to remain here during the winter, as I have a great deal to do here.

I find here a letter from my friend Dana, of the Exploring Expedition, dated Valparaiso. He seems not very well satisfied with his situation. I have not heard from any of you for a full year. Perhaps one of my sisters will favor me with a letter now that I am so near. Love to all.

CHAPTER IV.

A DECADE OF WORK AT HOME.

1840-1850.

ON Dr. Gray's return from Europe, the University of Michigan not yet needing his services, he settled in New York to work on the "Flora of North America."[122]

In 1841 he made his first journey to the mountains of North Carolina, of which he wrote an account in the "American Journal of Science" in the form of a letter to Sir William Hooker.

The country west of the Mississippi was just now opened to exploration, and for some years continued to afford an immense amount of new material to the botanist. Dr. Gray, and his friends Dr. Torrey and Dr. Engelmann especially, interested themselves in sending collectors with the various expeditious, explorations, boundary surveys, etc., and were kept very hard at work in studying and distributing the several collections as they came in. The difficulties of communication were great, postage was very dear, and the post-office rule that sheets, no matter of what size, could be sent as one letter, while the addition of any separate inclosure was utterly forbidden, added difficulties almost insurmountable to the transmission of any specimen. Even as late as 1850 the large parcels from St. Louis were sent by steamboat to New Orleans and then by sailing vessel to New York or Boston.

Foreign communication was not much better, as Dr. Gray writes to Sir William Hooker in March, 1840: "I have been waiting during the winter to write by some of the steamships, but they have disappointed us, and, though long expected, none reached us until the arrival of the Great Western a week or more since, which brought us fifty-six days' later intelligence from Europe."

TO W. J. HOOKER.

NEW YORK, May 30, 1840.

I have been tolerably industrious for some years, but have never labored as I have done this winter and spring. But I look now for a little respite, which I greatly need. I have this afternoon written the description of the last plant we have to give in the 1st volume of the "Flora" (a new cucurbitaceous genus, of which more anon); have prepared the last sheet for the press,—that is, of the work proper, which reaches to page 656 instead of 550, as intended; and have before me proofs of the supplement extending to page 672; what is yet to come will make up the volume to 720 pages! It has extended beyond all calculations or bounds, but we could not stop short. I hope to have done with the proofs early next week, when I expect to go immediately into the country and recruit for three or four weeks, for I am quite fagged out. Except, however, mere fatigue and the usual consequences of loss of rest, I was never, perhaps, more perfectly in health, and a fortnight or so of botanizing will restore my strength. You kindly inquire about my plans and prospects. These are so far favorable that they will give me (D. V.) another year of nearly undivided attention to the "Flora." Not long since I was officially informed that the opening of our university would be postponed another year, on account of unfavorable times, and the preparations not being sufficiently advanced. So I am told that I can have my time nearly all to myself until next spring (1841) if I wish (which of course I do), but without any salary, which, indeed, I could not with any propriety take while I perform no duty. By very close economy I think I shall get on for the year to come, and be able to accomplish a good deal of botanical work. I am going to pay the Michigan people a visit, and if they make good their promises made to me a year ago, as I have reason to think they will, their course towards me will have been liberal and honorable. I have good reason to hope they will eventually succeed in their plans.

By the London packet of the 15th of June we hope to send you and other friends some copies of the "Flora," parts 3 and 4. There are so many errors, so much bad printing, and so many things that we could now do much better, that I regret that any portion was published before my visit to Europe. Many of the most important corrections are given with additions, etc., in a supplement, but I hope we shall continue to improve as we go on. We can work to much greater advantage than before, from being much better supplied with books, as well as with specimens and information. Yet often do I wish to be within reach of your herbarium and library. Long accustomed to these advantages, you can scarcely appreciate the difficulties we often find. I was to-day wishing for a look at your Cucurbitaceæ; we have, as you know, but few of the order.

I shall not be able to visit Florida or any part of the Southern States this summer; indeed, I fear I shall be debarred from any botanical journeys for some years. I must direct all my time and strength to our "Flora." I hope we may complete another volume by the spring of next year. The way seems to be opening for increased facilities in sending a botanical collector to the Rocky Mountains. Our government is about to establish a line of military outposts quite up to the source of the Platte, in the principal pass of the mountains; and in a few years I doubt not we shall have small colonies in Oregon; but I know not when we shall be able to send a collector. I would like vastly to go after Grayia myself, but that cannot be at present. Nuttall has been giving a course of botanical lectures in Boston; and still remains there, I believe. My attempts to find Wilson's poem have not yet been successful. I shall esteem it a piece of good fortune if I succeed. I have engaged a friend of mine, a bookseller, also to search for it; and when I visit Philadelphia I shall inquire of some old people who knew Wilson. May God bless you, my dear friend; kindest, regards and affectionate sympathies to Lady Hooker.

Faithfully your attached,

A. GRAY.

TO ALPHONSE DE CANDOLLE.

NEW YORK, September 15, 1840.

MY DEAR FRIEND.... I had not forgotten our conversation on the subject of geographical botany. On my return I found I had a copy, a mere proof, of the little article I spoke of, and was about to offer it to you, but on examination it appeared to me much less important than I had supposed and perhaps led you to expect. But as it may be of some little use, I now beg you to accept it. I have added, here and there, the scientific names when the popular names only were mentioned.

The question you suggest as to the effect of the destruction of the forests on the climate is very interesting, and I think still unanswered. I fear it will be next to impossible to obtain data, even in this country, for its satisfactory determination. There are very few thermometrical observations on record of sufficient extent or exactness, except for the last eight or ten years. For a year or two I shall not be able to pay any attention to these subjects except to collect materials. But I am very desirous to afford you any aid in my power, and will attend to any suggestions you make, obtain any data which come in my way, or secure the services of our botanical correspondents scattered throughout our extended country. Pray tell me how I can aid you. The annual reports of the regents of the University of the State of New York are documents submitted annually to our legislature, and printed at their expense for public use. They relate chiefly to the condition of our colleges and higher schools, but for six or perhaps nine years past have also embodied the results of the meteorological observations made throughout the State under their instructions. The

"Reports" are not on sale, and the earlier numbers are not to be obtained except by some lucky chance....

The 3d and 4th parts of our "Flora," of which you speak so favorably, were sent to you through Baron Delessert, as I have already apprised you. By the time this work is completed we shall have settled somewhat accurately the geographical range of our plants, and have laid a good foundation for the comparison of our flora with that of other regions, etc. We shall soon begin to print the "Compositæ," and I trust in early spring we may see the second volume nearly or quite completed. Pray send me sometimes loose sheets of your articles or notices (those of your father and yourself) in the "Bibliothèque Universelle." I will sometimes translate them, if you do not object, or otherwise notice them, for the "American Journal of Science and Arts."

TO W. J. HOOKER.

NEW YORK, 15th January, 1841.

The dedication of the "Flora" we felt to be both a privilege and a duty; its favorable reception on your part gives us real pleasure.

I hope I have not offended Link by overstating his age. I am pretty sure I was so informed by Klotzsch who ought to know. You will now and then see some little articles or notices of mine in "Silliman's Journal." I prepare these notices merely to awaken and deepen the interest of our scattered botanists and lovers of plants, most of whom see that journal, and few of whom have any other means of knowing what is going on in the botanical world. We have, however, a few promising fellows who take the "Journal of Botany" or something of the kind. Should I have anything to communicate of interest to any other than our local botanists, I shall publish of course under my own name. You will receive with this a little notice of some European herbaria, which, commonplace as it must be on your side of the water, is useful to our own people. I have been as brief as I could, and have taken the pains to drop the first person singular. I am not sure but I have already sent you a copy through Mr. Pamphlin. Poor Rafinesque,[123] you know, perhaps, is dead; and I have attempted the somewhat ungracious task of giving some account of his botanical writings, which I will send you when printed.

I find that Townsend, Nuttall's companion, published, while I was abroad, an account of their journey. I have never seen a copy, and am told it is out of print; but I must try to find a copy for you. Townsend being poor, Nuttall waived his intention of publishing in his favor. I have heard that Townsend wishes to make a journey as collector of birds, plants, etc. I wish he would go to the southern Rocky Mountains, and trace them into

New Spain. Nuttall has brought home the Grayia. Have you ever received any more of Nuttall's plants, or has Boott? He is selling them to different persons for ten dollars per hundred; just such specimens as you received through Boott, or sometimes much better and more copious ones. I have some of his Compositæ in my hands, which Webb has ordered. He has a considerable number of Oregon and Californian Compositæ which Douglas did not get (and he failed to meet with many of Douglas's), and others in the States; as Pyrrocoma with rays. Nuttall ought to send all these to you.... I know with considerable accuracy what plants (Compositæ) are desiderata with you; and I will take the liberty of writing at once to Nuttall, and asking for such in your name. I shall ask for about one hundred Compositæ, and will extend the order to other plants if you desire it. He has, however, distributed nothing beyond Compositæ. Pray let me know at once if I have done rightly in this....

Among Drummond's Louisiana plants is the rarest of all United States Compositæ, Stokesia cyanea. It was pointed out to me by Arnott (January, 1839), but I have just examined Greene's specimens.

A. G.

NEW YORK, 20th May, 1841.

I have diligently labored about four months at Aster, in which, as I have after all not satisfied myself, I can scarcely hope to satisfy others; but I do think I have laid a foundation for the student of the species in their wild state. We had very copious materials, but could have done little in comparison without the aid of your collection, for which we cannot be too grateful. I am now occupied with Solidago, which is difficult enough, no doubt but not to be compared with Aster in this respect, partly because there are fewer species, and the synonymy much less involved, but chiefly because there are few in cultivation.

We rejoice to hear that Joseph and the Antarctic Expedition are getting on so well....

No further tidings of the steamship President! We have not until now surrendered all hope. One of the passengers, a stranger to me, but an acquaintance of a friend of mine, had charge of a small parcel for you, consisting chiefly of proof sheets.

October 15, 1841.

I will send by the next London packet (Quebec) and write more at [illegible] I have to-day sent on board that ship a box for Pamphlin, [illegible]g a parcel of plants for you (all of any consequence of my small [illegible]llection with some others). Few as they are, I trust it will give me a pleasure I seldom can enjoy—that of adding something to your herbarium. Mr. Brydges takes also for you the proofs of a gossiping article on the botany of the southern Alleghanies, etc., which I have taken the liberty to address to you, and hope it will meet your approval. I shall send you clean copies, as soon as they are printed. The article will not appear here until the 1st of January. I send you also some ripe seeds of Diphylleia for your garden. I have live roots in the care of a cultivator. If they live shall send you one in the spring....

I must not forget to mention that my package also comprises a set of Ohio Mosses from my friend Sullivant, of whom I have often spoken, and of whom as a botanist we have high hopes, as he has an independence (for this country), talent, and much zeal. If not too much trouble, I join with him in requesting you to name them according to the numbers, by which you will do him great service, as he designs to study and collect American Musci especially.

TO GEORGE ENGELMANN.

NEW YORK, November 30, 1841.

DEAR DOCTOR,—Don't hesitate about sending me anything for fear I may already have it. Very many plants pass through my hands while I am describing, but my own herbarium is not very rich; and duplicates will not oppress me. Mr. Carey does not keep European plants except those identical, or supposed identical, with North American species. Browne, however, does, and I dare say would be glad to have any you can give him. They are the gentlemen mentioned in the "Flora." ...

Eupatorium Engelmannianum, sp. nov. Am. Bor., semina misit Engelmann. Can this be it, think you? If so pray help me to it; and to anything else you can, as I mean to give addenda et corrigenda to the Compositæ at the end of the order, if I ever get through this formidable job. No wonder seven years' labor at them ruined De Candolle's health. You know he is dead? He died the 9th or 10th of September last....

I send you my article in the January number of "Silliman's Journal" with a little one by Sullivant,—by mail. I am extremely busy this winter, but I hope always to answer your letters promptly, and to attend to your desires as well as I can, whence I beg you to continue your useful correspondence.

March 30, 1842.

It is not a great while since I got all the copy ready for the number of the "Flora" now printing,—during which I could do little else. Immediately this was done I completed an arrangement with my publishers for preparing a handsomely got up Introduction or Text-Book of Botany, for schools, lectures, private students (medical, etc.), which must be out on the 1st of May next. Owing to illness I have as yet written almost nothing, and besides have to superintend all the drawings, as they must be made by a person unacquainted with botany; and at the same time I have to correct the proofs of about thirteen sheets yet of the "Flora," so that I am almost distracted when I think how I am to accomplish it here, where I have to see personally to almost every detail. But I must do it, as I hope to lay the foundation for a popular and—what is of consequence to me—a profitable work.

TO W. J. HOOKER.

NEW YORK, 30th March, 1842.

The last steamship left Boston so soon after I received your kind letter that I was unable to answer it by that conveyance. I intended to send this by the Columbia steamer of the 2d prox.; but I learn that having broken her shaft in the outward voyage she is to sail back to England; when it comes to canvas I have more confidence in our old liners, and therefore send by New York packet.

Have you not seen or heard of Nuttall yet? He sailed for England on Christmas last, to take possession of property left him by some deceased relatives.

I should not feel a residence in Michigan as a banishment. I am fond of a country life. But at present I see almost no hopes of usefulness there. Like all our new, and some of our old States, they have squandered the means they once possessed and encumbered themselves almost irretrievably with debt. On my return from Europe in the autumn of 1839, I received a letter stating that they had nothing yet for me to do, and permitting me to spend the winter in New York. In the spring of 1840, a committee of the regents wrote to me, to relinquish the provisional salary (of fifteen hundred dollars, on which I had been placed) for one year from that date, they relinquishing my services for that period and allowing me to devote my time to the "Flora," etc. I at once accepted their proposal; but although another year has now elapsed since the expiration of the period to which they proposed to limit this agreement, not a word have I heard officially or unofficially from Michigan. I have quietly awaited the result, ready at any moment to obey their call; but having no income for the last two years, I have been greatly embarrassed, and have struggled through great difficulties, I scarcely know how. Notwithstanding, I have thought until recently that I ought not to seek any other situation. I shall now write to Michigan immediately, inquiring whether, in their present condition, they are ready to fulfill their engagements with me, or whether they would prefer to accept my resignation, which I shall offer. I expect, and on the whole hope, they will accept it.

In December, or nearly the 1st of January last, a friend of mine here, who had some casual conversation with the President of Harvard University, wished me to let my name be known as a candidate for the vacant chair of natural history there. After reflecting for a week or two, I wrote to B. D. Greene[124] for some information on the subject, saying that, if freed from other engagements, I would like the botanical part of the professorship, but not the zoölogy: and that the former, with the charge and the renovation of the Botanic Garden, would be quite enough for one.

In January I made a flying visit to Boston, where I had never been, and knew no one personally but Greene, to whom, and to Professor Bigelow,[125] I expressed my views; but we none of us expected that anything would be done at present. I incidentally learned, however, not long since, that the men of science would generally be well pleased to have me at Boston, and that some with whom I had almost no acquaintance were using their influence to that end. I was never more surprised, however, than this very evening, when I received from President Quincy an official letter, offering me the professorship provisionally, with a small salary, to be sure, for the present, but with only the duties of the botanical portion.

The president states that the endowment is $30,000, yielding an income of $1,500, which, however, not being adequate to constitute a full professor's salary on a permanent foundation, the corporation deem it both their duty and the interest of the professorship to continue for a few years, in a modified form, the policy they have hitherto pursued, and by applying one third of the income annually to the augmentation of the capital, enable themselves to place the professor of natural history, at no distant period, on an equal footing with the other professors of the university. "To this end they propose to limit your duties, in case you are willing to accept the professorship, to instruction and lecturing in botany, and to the superintendence generally of the Botanic Garden (which they wish to renovate); limiting for the present your annual salary to one thousand dollars;" thus enabling me, as the communication proceeds to say, to devote all my time at present to my favorite pursuit, and to go on with the labors I have in hand. I have reason to hope, also, that by the time they are ready to give me the full salary, the zoölogical part will be separated from the professorship, with a distinct endowment. The Botanic Garden has an endowment of $20,000. If I should take this place, I should hope to see it better endowed before long, and should immediately set about the introduction of all the hardy trees and shrubs,—and indeed to enrich it as fast as possible with all the American and other plants that could be procured. In that case, separated from yourself by only fourteen to eighteen days' navigation, I could hope to be a useful correspondent to you at Kew, and to show my gratitude for your continued kindness to me. I must here conclude, by stating that the president's letter to me is to be deemed confidential, in case I do not accept the offer. I must therefore beg you to consider this letter likewise confidential, until you hear further from me, which you may expect to do as soon as anything is settled in regard to this matter. I am the less reluctant to leave New York since our good friend Dr. Torrey is at Princeton, New Jersey (only four hours from New York), renting his house in town, where for the present he will only remain during the winter. We have worked so long together that I shall feel the separation greatly.

NEW YORK, 30th May, 1842.

I have the pleasure to inform you that having accepted the offer from Harvard University of which I apprised you in my letter of April 1, I was appointed to the professorship on the 30th of April last. The incessant occupation of this month has prevented me from writing to you sooner, and still prevents me sending anything beyond this hasty note. I hope in a week or so to have my new text-book finished, when I shall visit Cambridge to make the necessary arrangements for my removal thither. I hope hereafter to be a useful correspondent to you, in the way of supplying you with seeds and living plants of our own country, and when I see what can be done with our Garden I shall probably ask you to aid us. I wish to visit the mountains of Carolina again, in autumn, to procure roots and seeds....

In the spring of 1842, as his last letter intimated, Dr. Gray was appointed to the Fisher professorship of natural history in Harvard College. He was then thirty-one years old. He removed to Cambridge in July, taking lodgings near the colleges at Deacon Munroe's, on what is now James Street.

Before Dr. Gray came to Cambridge he had been elected into the American Academy (November 10, 1841). He threw himself with the greatest interest into its work. Scarcely any winter storm kept him from its meetings; all other engagements had to give way. And when new life began in its publications, many of his most important papers appeared in its volumes.

He was also influential in establishing a scientific club consisting of members of the college faculty and

other friends in Cambridge. Of this, too, he was a most faithful member. The club met twice a month at the houses of the different members in turn, and the one at whose house it met was expected to bring forward some subject, generally from his specialty, which later was discussed and criticised. Many of the new interests in science were here first presented by Dr. Gray.

Among the founders and early members were, Charles Beck, Francis Bowen, Admiral Davis, Epes S. Dixwell, Edward Everett, President Felton, Asa Gray, Simon Greenleaf, Thaddeus Mason Harris, Joseph Lovering, Benjamin Peirce, Josiah Quincy, Jared Sparks, Daniel Treadwell, James Walker, Joseph E. Worcester, the lexicographer, and Morrill Wyman, M. D. Later, among those no longer living, were added at different times Louis Agassiz, Thomas Hill, Joel Parker, Emory Washburn, and Joseph Winlock. The club is still in existence.

TO JOHN TORREY.

BOSTON, Monday, 25th July, 1842.

MY DEAR DOCTOR,—Having time before the mail closes to write a harried letter, I hasten to let you know that I have this morning secured lodgings at Cambridge, at a retired house, off the main road, about halfway between the colleges and the Garden. For $3.00 per week, I have two rooms, one pretty large, one moderate (of which I shall make a bedroom), a small nearly dark bedroom which I shall shelve and use for my herbarium, and three closets, furnished decently (but not extravagantly!!), in a house where there can at most be only one other lodger, and he must ascend by a different staircase from mine,—the rooms and bed linen, etc., to be kept in order.

I am to board at an adjacent house, to which I have access by a private gate through the garden. The latter house belongs to Mrs. Peck (widow of my predecessor), who boards there, and who I see has bestirred herself to contrive and effect this arrangement. I am to take possession next Monday. Meanwhile I am Mr. Greene's guest here, where I have the house for the most part to myself. I arrived here Friday morning, just in time to miss the president, who had just started for Portland, and has not yet returned. I have seen Bigelow, Emerson,[126] etc., and have been looking about among the libraries here, and endeavoring to arrange matters so as to procure just, and only such, books for the college as are wanting. I am pleased to find a complete copy of "Linnæa" at the library of the American Academy.

I passed last Sunday all alone in Greene's house. Mr. Emerson met me coming from Park Street Church, and on telling him that I was of Orthodox faith, he said he was very glad of it, although not altogether of that way himself.

I have been only twice to Cambridge, whence I have just returned, and where you may address your letters. But I can do little there until the president returns, by which time, however, I must trust to have my list of books ready. I have just written to Mr. Wiley to send on my boxes, and hope next week to get nearly in working order. I now think of remaining here (studying Compositæ, etc.) through the month of August, and then visiting Mt. Washington, if I can get money and a companion (I shall ask Oakes), and in September going (via New York?) to western New York, where I wish to collect roots and seeds as extensively as may be. I will soon make out a list of some things I would like Knieskern to get for me in the pine barrens.

Tell E., also, that I must write her about a learned lady in these parts, who assists her husband in his school, and who hears the boys' recitations in Greek and geometry at the ironing-board, while she is smoothing their shirts and jackets! reads German authors while she is stirring her pudding, and has a Hebrew book before her, when knitting [? netting—A. G.]. There's nothing like down East for learned women. Why, even the factory-girls at Lowell edit entirely a magazine, which an excellent judge told me has many better-written articles than the "North American Review." Some of them, having fitted their brothers for college at home, come to Lowell to earn money enough to send them through!! Vivent les femmes. There will be no use for men in this region, presently. Even my own occupation may soon be gone; for I am told that Mrs. Ripley (the learned lady aforesaid) is the best botanist of the country round. But the mail is about to close; this nasty steel pen refuses to write; dinner is ready, and so with love to all, I subscribe myself,

Yours most affectionately,
A. GRAY.

TO W. J. HOOKER.

CAMBRIDGE, 30th July, 1842.

MY DEAR SIR WILLIAM,—It is indeed a long time since I have heard from you; although, indeed, I can well suppose that, in your new situation,[127] you are too much occupied to write frequently to your friends on this side of the ocean. Having finished my little "Botanical Text-Book" (a copy of which is sent you through the publishers, Wiley & Putnam, who have an office in Stationer's Court, Paternoster Row), and packed up my things at New York, I have just taken possession of my situation at Cambridge. The Botanic Garden, which has a good location, contains over seven acres of land, and the trees have well grown up. It already contains some good American plants, and I shall immediately commence a plan of operations with the view of accumulating here, as fast as possible, the phænogamous plants, etc., of the United States and Canada; and hope to supply you with such of our indigenous species as you may desire. I wish I could know what plants are likely to be acceptable to you, that I may not send you what you already have. I must postpone to next year my contemplated visit to the mountains of Carolina, where I can make a fine collection of interesting plants for cultivation. Perhaps I can also visit Labrador next year. This autumn I must confine myself to an excursion to the White Mountains, to the western part of New York, and to the pine barrens of New Jersey. I shall most gladly share the seeds and roots I collect with you. My good friend Mr. Sullivant, also promises me the living Sullivantia and many other interesting plants.

Let me also say, my dear sir, that any duplicates you can spare us from your noble institution will be truly acceptable and in the highest degree useful to us, as we have very few exotics and hot-house plants. We have a good gardener, and I think I can promise you that whatever you choose to give us shall be sedulously taken care of.

Dr. Torrey is now at Princeton. I had the pleasure of spending a week with him not long since, and hope to visit him again early in the autumn. I shall miss him very much. I am here more favorably situated with respect to books than at New York. I hope next week to begin again with the "Flora," and perhaps to finish the Monopetalæ.

TO GEORGE ENGELMANN.

CAMBRIDGE, 26th July, 1842.

MY DEAR DOCTOR,—I hope to get settled here, and in working order in a week or so; to work at Compositæ, all next month, and to occupy a part of September and October in collecting the roots and seeds of plants, of the White Mountains, of western New York, etc., for our Botanic Garden here; which I wish to renovate, to make creditable to the country and subservient to the advancement of our favorite science. I wish to see growing here all the hardy and half-hardy plants of the United States (as well as many exotics, etc.), and shall exert myself strenuously for their introduction. The Garden contains seven acres; the trees and shrubs are well grown up; we are free from debt, and have a small fund. The people and the corporation are anxious that we should do something, and I trust will second our efforts.

Allow me therefore to say that yourself and your friend Lindheimer[128] in Texas would render me, and also the cause of botany in this country, the greatest aid (which I will take every opportunity of publicly acknowledging), if you will send me roots or seeds of any Western plants, especially the rarer, and those not yet figured or cultivated abroad. But nothing peculiar to the West and South will come amiss. I am calling on all my correspondents to assist me in this matter; which, by giving me the opportunity of examining so many living plants, will vastly increase the correctness of our "Flora." I shall not be idle myself. I will defray all expenses of collection and transportation (boxes may be sent via New Orleans, directly to me at Boston). If you wish to cultivate anything that I have or can procure, it shall be forthcoming. Pray let me hear from you on this subject.

TO JOHN TORREY.

CAMBRIDGE, 15th September, 1842.

MY DEAR FRIEND,—Your letter of the 6th inst. awaited my return from the White Mountains last evening, and I must drop you a hasty reply by this day's mail. I started for the mountains almost at a moment's warning. Emerson, who was to accompany me, being called down to Maine, wrote me unexpectedly to meet him on Monday or Tuesday of last week at the Notch. I had just time to look up Tuckerman,[129] the very morning of his arrival! and to get his consent to meet me on Monday morning at the cars for Dover. Monday evening we reached Conway, New Hampshire, thirty miles from the White Mountains (full in sight); and Tuesday, in a one-horse wagon, we reached and botanized up the Notch to Crawford's at its head. Emerson had been there, and returned to his father's in Maine, having learned his brother's arrival from France in the ship that brought Tuckerman. We made two ascents to the higher mountains; slept out one night; cold weather; a good deal of rain, but had some very fine weather for views. We saw the ocean distinctly, which is only possible under favorable circumstances. I made a fine collection of living plants, which was the chief object. Although too late for botanizing, yet I got many good alpines in fruit, some few in flower. When I see you, which I trust will be soon, I will tell you particulars, and bring specimens of the few plants collected that will be needed in your herbarium.

I have seen the president this morning, and find that Mr. Lowell has returned, but all are so busy that I doubt if they will settle anything about our affairs until the last of next week. Consequently I shall be kept here all next week. I shall immediately, at Mr. Quincy's desire, or rather approval of my intimation, draw up a plan of my wishes for the management of the Garden, and shall ask for a specific appropriation, of small amount, for obtaining live plants, paying bills of transportation, etc. If I succeed, I may then be able to engage Knieskern to procure some New Jersey plants, as well as go to western New York myself; but I fear this delay, with the advancing season, will perhaps prevent the latter.

Saturday afternoon, 5th December, 1842.

The parcel of Compositæ, etc., of the Far West has only just come in. I have looked over the Compositæ with some excitement. Some few new and the old help out Nuttall's scraps, etc., very well. Tetradymias this side of the Rocky Mountains!! Some new Senecios, especially, from the mountains, near the snow line. How I would like to botanize up there!...

I wish we had a collector to go with Frémont. It is a great chance. If none are to be had, Lieutenant F. must be indoctrinated, and taught to collect both dried specimens and seeds. Tell him he shall be immortalized by having the 999th Senecio called S. Fremontii; that's *poz.*, for he has at least two new ones....

I have the privilege of expending one hundred dollars in botanical illustrations,—to be the property of the college and to be increased from time to time. How do you advise me to proceed in the matter?

Though greatly behindhand, I must get Compositæ all done this month. Then if you could have the Lobelias and Campanulas ready, I think we could print the latter part of January, and I get everything off my mind and ready for teaching 1st of March....

This letter you see has no beginning, as I have scribbled down memoranda for a day or two past, as they occurred to me. I am deep among Thistles, which are thorny (though I see that they are satisfactionable, all but one little group of two or three species), and have been considerably interrupted, or I should have written you sooner.

TO MRS. TORREY.

CAMBRIDGE, Wednesday evening, December 14, 1842.

It is some time since I have written to Princeton, and longer since I have heard from any of you; for I believe you are every one in my debt. This, however, has not restrained me from writing, and I have only waited until a proposition very unexpectedly made me a few days ago should be disposed of. I have been invited to lecture before the Lowell Institute next year, and have had the hardihood to accept! A celebrated lawyer here says that he never hesitates to take any case that offers, to be argued six months hence! I have taken this in much the same way. But when the time draws near I dare say I shall call myself a very great fool. But it is now neck or nothing. The money will be really very useful to me; to decline the offer, coming from one of the most influential of the corporation of the college, would have had an unfavorable effect on my prospects, which moderate success will greatly advance. The pay is $1,000 for twelve lectures, or $1,200 if they are repeated in the afternoons. Instead of the latter, I have proposed to give a collateral, more scientific course of about twenty lectures, with a small ticket-fee to render the audience more select, and for which I should get about $500, making $1,500 in all. The Institute will pay for full illustrations. Mr. Lowell offered at once to engage me for two or three years; but I told him he had best wait to see how I succeeded. Mr. Lowell told me that he was in treaty with two of the most distinguished orthodox divines in this country for courses on Natural Theology and the Evidences of Christianity; the one to commence next year, the other the year after. I do not doubt one is President Wayland. Who can the other be? Tell Dr. Torrey he hopes to get Faraday next year; and Mr. Owen the year after.

I should not wonder if my appointment were in some degree owing to a little piece of generosity in a small way that I played off not long since. The president has once or twice asked me to hear the Freshmen next term in a course of recitations from a text-book on general natural history as a matter of favor, as he did not wish Mr. Harris or any one else to perform this duty; and offering me, of course, additional compensation, I suppose $200 or so. I found, however, that this pay would come from the funds of the Garden, let who would perform the duty. So to prevent that, I offered to perform the duty, but to receive no pay for it. At the same time, however, I got the corporation to appropriate $100 for illustrative botanical drawings, which otherwise would have come out of my own pocket. So you see I have work enough ahead, if I live, to give me both occupation and anxiety. I have been driving away at the "Flora," of late, very hard, hoping to come to New York to print next month; when all this matter must be laid aside, and I must prepare for my lectures, etc., for next term, which commences about the first of March.

I am very tired, having been in Boston all day,—at tea at Mr. Albro's, our good pastor, where I met Mr. Dana, father of "Two Years before the Mast" Dana, and passed the rest of the evening at Professor Peirce's.[130] To-morrow I hope to have for study; but the next day I shall be obliged to go again to Boston, and perhaps stay till evening for a soirée at Mr. Ticknor's.

The Latimer case has greatly increased the abolition feeling in this State, besides showing that the recent decision of the Supreme Court will in fact operate in favor of the runaway slave. It is not probable that another slave will ever be again captured in Massachusetts. There is a petition to Congress in circulation, designed simply to express the feelings of Massachusetts, which will probably be signed by almost every person in the State.

TO JOHN TORREY.

CAMBRIDGE, January 3, 1843.

Your letter, truly welcome after so long an interval, reached me yesterday. I should have been very glad to be with you during the holidays, but cannot think of leaving before I finish these interminable Compositæ. I hoped to have accomplished this on Saturday last; all but taking up some dropped stitches; but was a good deal interrupted last week. The December number of "Annals and Magazine of Natural History" (of which Professor Balfour is the botanical editor) contains a very complimentary notice of the "Botanical Text-Book," accompanied with a few judicious selections, which shows that the writer has looked it over carefully; and winds up by terming it the best elementary treatise (as to structural botany) in the English language. So easy is it to get praise where it is not particularly deserved!...

My great object for next year is to attempt to raise $10,000 from some of our rich men, to rebuild our greenhouse on a larger and handsome scale. There are a few men, who have never given anything to the college, who may perhaps be induced to give for this object.

TO GEORGE ENGELMANN.

CAMBRIDGE, MASS., February 13, 1843.

I note with interest what you propose in regard to Lindheimer's collections for sale in Centuriæ, fall into your plans, and will advertise in "Silliman's Journal" (and in "London Journal of Botany") when all is arranged. Pray let him get roots and seeds for me. I will do all I can for him. But if the Oregon bill passes, a party under Lieutenant Frémont, or some one else, will go through the Rocky Mountains to Oregon; and parties of emigrants or explorers will go also. Now why not send Lindheimer in some of these? Probably the government party would afford him protection, and probably he might be formally attached to the party. Frémont will not take Geyer;[131] but I believe he wants some one. The interesting region (the most so in the world) is the high Rocky Mountains about the sources of the Platte, and thence south. I will warrant ten dollars per hundred for every decent specimen. If he collects in Texas, eight dollars per hundred is enough. I write in haste, hoping this plan may strike you favorably and be found practicable. Let me know at once. The opportunity should not be lost. Do send Lindheimer to the Rocky Mountains if possible.

TO W. J. HOOKER.

CAMBRIDGE, February 28, 1843.

I found your most welcome letter on my return from New York a few weeks since, and have since sent it to Dr. Torrey, who was equally delighted with myself at the opportunity of hearing from you.

Our term opens to-day, and I am just on the point of commencing my course of botanical lectures, which is rather formidable to a beginner. So you will excuse my hasty letter. I would not miss to-morrow's steamer, as I wish to say that your offer to furnish our Garden—the great object of my care—with hardy plants from your rich stores at Kew delights me much. I have only to say that everything you can send will be truly welcome. Our stock of European hardy plants (whether herbs or shrubs) is small, and consists of the commonest and oldest-fashioned things in cultivation. These, and every Californian, Oregon, and Texan plant of which you have duplicates to spare us (or seeds), whether hardy or not,—these are the plants I am just now most desirous to accumulate. Greenhouse plants are scarcely less welcome, but of those I will write more particularly hereafter. Can you send us a young Araucaria imbricata and Stuartia pentagyna?

My plans for accumulating American plants were put in operation too late last autumn to give us much as yet, but my correspondents throughout the country seem interested in the matter; some will reach me this spring, and still more, I trust, in the autumn. With regard to all these, as soon as I see them growing, so that I can send them with authentic names, I shall most gladly share with you.... I shall continue to direct all my energies to the advancement of our amiable science in this country, not, I trust, in vain. I have a plan to publish, from time to time, figures of rare or interesting North American plants, chiefly those cultivated in our Garden and those upon which I may throw some light. I think there are persons enough here interested in the matter, including gentlemen of public spirit here, who would encourage it for the Garden's sake, to nearly defray the expense, which is all I desire or expect....

What a charming place you must be making of Kew! What a field for the botanist!

TO MRS. TORREY.

Thursday evening, 2d March, 1843.

You will be anxious to hear how my first lecture succeeded, knowing it was to have been given to-day.[132] But you must wait a week longer. Since my last letter was dispatched the president, finding the class would hardly be ready, desired me merely to meet them to-day for the purpose of pointing out the subject in the "Text-Book," arranging general plan and all that, postponing my lecture to Thursday of next week. This I was most ready to do, as it gave me the opportunity of entering by degrees upon my task, feeling my way instead of making a plunge in regular desperation. The great thing is self-possession. The moment I get that I shall feel tolerably safe. So I met my class to-day, arranged matters, and made a few remarks without stammering a bit, so far as I recollect, or speaking much too fast. My class consists of about two dozen students (undergraduates), mostly Seniors, besides which any law or divinity students and resident graduates who choose can attend, and several probably will. For my recitations in natural history generally, I have divided the Freshmen into four sections, about sixteen in each, two of which I meet on Fridays, and two on Tuesdays; have given them their lessons, and to-morrow, consequently, I commence these recitations. I must not forget to tell you that since my return the Sunday-school class left by one of our people who has removed to Boston has been given me, a class of eight or nine very intelligent misses, varying from sixteen years old to twelve, all of one family, though originally of three, some being sister's children (orphans, etc.). I am greatly pleased with them, delighted with their docility and intelligence, and anticipate a very happy time. So you see I have three sets of scholars, on different subjects. I ought to be "apt to teach."

Saturday morning.—I must dispatch my letter by to-day's mail, and as I am going to Boston, where I have not been for a week, I will drop it in the post office there, to insure its transmission by this afternoon's mail.

Yesterday afternoon I met the first two sections of my class of Freshmen for recitation. It went off very well. I am pretty good at asking questions. The lads were well prepared. Next Tuesday I meet the third and fourth sections; and on Thursday, the ides of March, I give my first lecture on Botany. If I succeed well, I am sure no one will be more pleased and gratified than yourself, and that of itself is enough to incite me to effort. If I don't altogether succeed, neither satisfying myself nor others, I shall not be discouraged, but try again, as I am determined to succeed in the long run. Nil desperandum. I shall have the president to hear me; but he is said always to fall asleep on such occasions, and to be very commendatory when he awakes.

I now board with the sister of my landlord, Deacon Munroe, a table of only five, one professor, one tutor, and two advanced law students. We yesterday commenced the experiment of dining at five o'clock, much to my gratification, and if the other gentlemen like it as well as I do, we shall continue to dine at that hour, until summer at least. It is very cold here; though the sun shines brightly all day, it scarcely thaws at midday.

CAMBRIDGE, March 18, 1843.

Your most welcome and long-expected letter of the 14th reached me only this noon. This first day of leisure of this week has been a very busy one. I have been to town, and just got back. I have had to work very hard this week. I have got my course of recitations for the Freshmen on Smellie well in progress, and am quite interested in it, though at first I thought it would have been a great bore. The class are generally very much interested, and give promise that I shall reap the fruits of my labor when they become Sophomores or Seniors and attend the botanical lectures, for which I think I am laying a foundation. I am now perfectly at ease in my mode of teaching them; I am pretty good at questioning, and I give them plenty of illustration, explanation, and ideas not in the book, which pleases and interests them. In one of the divisions last week, while giving them a sort of lecture, two hours long! (to which they listened well; for I gave them, or those who chose, the opportunity of going at the expiration of the regular hour, but not one of them budged), turning my head at a fortunate moment, I caught one of the fellows (rather a stupid fellow, a boarder with me last term) throwing his cap to his companion or playing some trick. You know I can scold. So I gave him about half a dozen words that made him open his eyes wide; and I do not think that he, nor any of that division, will venture upon anything of the kind again very soon.

As to the botanical class, which now numbers thirty-seven, I have given two more lectures, for I lectured both Thursday and Friday, on the last occasion, which was a sort of recapitulation quite without notes, as a trial. I am convinced that for lectures with much illustration I must have only heads and leading ideas written; for others, I will write nearly in full. I saw Miss Lowell ... the day before my first lecture, and promised to call upon her very soon if I succeeded well. Meeting her the other evening at Professor Sparks's, she reproved me for not keeping my word. I very honestly and sincerely replied that I had not succeeded well, and was waiting until I was better satisfied. Quite to my surprise, I found that the class, at least those she had seen, her great-nephew and others, were well pleased with it. I will not repeat their expressions, as retailed to me by Miss Lowell, because I cannot but suspect that young Lowell may have been trying to humbug her. I feel I have so far acquitted myself very poorly as a lecturer; but I am sustained by the firm conviction that I shall in the end do very well, for a common college class.

TO JOHN TORREY.

May, 1843.

I have been speaking about the bones of the Zygodon, and there is a disposition to get up a subscription in the Natural History Society and buy them, if still for sale, the price not too great, and if Dr. Wyman, on seeing them, recommends the purchase. Do you know the price? And whether they can still be seen in New York, at Carey's storehouse? The Boston zoölogists are far from praising De Kay's Report. I heard Silliman on electro-magnetism the other evening (which hardly belongs to chemistry): great show of experiments; lauded Henry finely. He is finishing off with galvanic deflagration. Will Frémont go west this year? So Mr. Carey is going to Buffalo. Occupation will be the best thing for him; but we shall miss him in New York....

Monday afternoon, 9th May.

I have a few of Frémont's plants up from seeds. The two pine-trees and the Pyxidanthera were received in good condition, to my great wonderment. Pyxidanthera is in full bloom, and a drawing of it nearly finished (as well as of Oakesia, about which I have some new matters that are curious) by the eldest Miss Quincy, whom I have pressed into the service....

Rhododendron Lapponicum, from the White Mountains, is just bursting into flower. I am building rock-work, but we get on slowly. All the work of the Garden comes together this spring, and all in a heap.

TO W. J. HOOKER.

CAMBRIDGE, 30th May, 1843.

... The community here are very liberal and public-spirited. They have just given by subscription $25,000 for a telescope, etc., for our observatory. The college have given me the use of seven or eight acres of land lying around the observatory, finely situated and diagonally opposite the Botanic Garden, as an addition.[133]

As soon as our garden begins to increase and prosper, I hope in a year from this we shall attempt (and doubtless succeed) in raising the funds for a new conservatory, hot-house, etc.

TO GEORGE ENGELMANN.

CAMBRIDGE, 22d June, 1843.

When you get sufficient collections from any of these botanists for distribution, you will please forward me a set, with your own critical remarks. Although I excessively dislike to study special collections far ahead of my work, yet in these cases it will be important, and I will consent to do it. If I thus join in the responsibility and labor, which will be great to a person with his hands so full as mine, the articles written on the subject and the new species must bear our joint names.

You cannot have failed to perceive that the genus Astragalus is not well done in the "Flora." ...

I agree with you generally in the impropriety of too much multiplying names of species after the collectors, etc., yet I think these are good names, easily remembered, and particularly advisable in very large genera. My practical rule is to name such species after the discoverer, etc., if I cannot find any really pertinent characteristic name unoccupied....

There is much to be done, and so little time that I often wish I could divide myself into a dozen men, and thus get on faster. Let us, however, take particular pains to do everything thoroughly as far as we go.

TO MRS. TORREY.

CAMBRIDGE, July 22, 1843.

I find Cambridge, in vacation, as quiet as possible,—most people away. The president's family were at home, and unaffectedly glad to see me; but several of them, including Miss Susan, who makes drawings for me, are about to set out on Monday for Lake Champlain, Montreal, and Quebec; to be absent nearly to the time that I hope to leave here again; for I find, from the way the president takes it up, that I shall have no difficulty in obtaining the sanction of the corporation to my proposed mountain tour. But of that I shall know certainly in a day or two. In that case I shall hope to see you again in the latter part of August, perhaps as soon as the middle....

Dr.—— came here the day I returned. He still garnishes, as ever, his lack of ideas with a deliberate profundity of words.

I found on my return a letter from my brother, announcing the approaching marriage of my youngest sister; which event took place, I suppose, on the 20th inst., the day I left New York. Had I received the letter in New York, I should have arranged to be present on the occasion. I wonder if my turn will ever come!

TO W. J. HOOKER.

CAMBRIDGE, 11th August, 1843.

I leave home this afternoon for New York, on my way to the Alleghany Mountains in the north of Virginia, where I expect to meet my excellent friend Mr. Sullivant, of Ohio. We hope to trace the more westerly ranges of the mountains down to North Carolina and Tennessee, to revisit my old ground in Ashe County, etc., and to continue our journey farther south into Georgia, coming out at Augusta on the Savannah River; thence I may go to Charleston and return by water. But if time allows I shall perhaps run through upper Georgia and Alabama, to the Tennessee River, down that to the Ohio, and thence home. My chief object is to obtain live plants and seeds; we shall be too late in the season for the best botanizing, yet I think we shall be in the best time for Compositæ. Mr. Sullivant will turn his attention primarily to the Musci; but we shall let nothing escape. Thus at last I may hope to be somewhat useful to you as a correspondent for your Garden.

I learn within a few days that Ross's expedition has been heard of from Rio. Doubtless Joseph will have reached home before this letter arrives, and I may congratulate him—and yourself—upon his most gratifying success, which has laid a broad and sure foundation for his scientific eminence. His Flora Antarctica must be of the very highest interest and importance.

TO JOHN TORREY.

ASHEVILLE, Saturday, September 30th, 1843.

MY DEAR FRIEND,—Your two letters which awaited my arrival—the one at Jefferson, the other at Asheville—were indeed refreshing. Our long journey through Virginia brought us behind our estimated time, and hurried the later and more interesting part of our operations; for Sullivant was getting very impatient, as I wrote in my last, just as we were hurrying away from Jefferson.

I doubt if I got anything of much interest in Virginia, except Buckley's (and Nuttall's) Andromeda, Rhamnus parvifolius on the waters of Greenbrier, (where did Pursh get it?), Heuchera pubescens in fruit and Heuehera hispida Pursh!! out of flower and fruit, so that I detected it by the leaves only (and got good roots), not far from where Pursh discovered it, but more west, on the frontiers of a range of mountains where this very local species doubtless abounds.

From Jefferson went to Grandfather; had a fine time and good weather; explored the old fellow thoroughly, but found 140 new Phænogams. Sullivant made a great haul of Mosses and Jungermanniæ. Found the Moodys heartily glad to see us. The elder brother is married since our former visit. Miss Nancy delighted with the calico dress I brought her, and made me promise to ask some of my lady friends at home to cut out a pattern for her in newspaper and send by mail,—to be in tiptop style,—in the very height of the fashion! Poor Miss Nancy! How she would look! The "old gentleman" (Mr. Carey) was most affectionately inquired after. Indeed Miss Nancy is perfectly in love with him, and sacredly keeps the sperm-candle-end he gave her as a relic. She gave me a most amusing account of the wonderment which our visit caused. To it she attributes the advantages they now enjoy both for religious and secular instruction. For we found a young Episcopal clergyman, sent by the bishop, resident in the neighborhood, where he has spent already almost a year,—a perfect hermit, so far as civilized society goes. Yet he is busily occupied, and nearly contented, has built a little cabin in full view of the Gothic Grandfather, and I hope is doing much good. He accompanied us to the mountain, but did not remain over night in our encampment, having a distant service on Saturday. His name is Prout. Mrs. Torrey will remember something about his history, which will in part account for his willingness to spend a few years in this solitary region. I had hoped to hear him preach on the Sunday we passed at the Moodys' on our return from the mountain; but he preached at a station ten miles off.

A. GRAY.

In one of his later mountain journeys Dr. Gray passed again through Val Crucis in June, 1879; and the following extract from Mrs. Gray's journal gives the sad fate of the little mission colony.

"In the afternoon we came upon Val Crucis.... It seems, years ago (in 1841) when Dr. Gray, Mr. John Carey, and others came exploring in the mountains, Mr. Carey was laid up for a while in a farm-house, and talking with the good people found them woefully ignorant, especially of everything relating to Christianity. So when he went back to New York he corresponded with the Southern bishop, who bestirred himself, and a mission was sent into the mountains. They settled at Val Crucis, and so named it. It was in the early days of Ritualism, and the young men thought to found something like the early monastic settlements in England, and as it seemed to the ignorant people, played strange pranks and preached wonderful and incomprehensible doctrines which puzzled and bewildered them; then Bishop Ives went over to the Catholic Church, and it all died out; and here is the church (the rude timber church), with still a few members, but all the farms and settlements passed into other hands—as far as I could make out into the hands of a rich old man, who lives anything but a holy life, and whose boarding-house for the saw-mill hands in Val Crucis is an awful degradation! I saw at the Duggers a large old Bible, and on it printed 'Society of the Holy Cross, Val Crucis,' which the children were using to paste stories and pictures in!"

The journal continues:—

Monday and Tuesday.—Crossed the Blue Ridge, descended John's River, and went to near the base of Table Mountain. Wednesday, ascended it. Was fortunate enough to get Hudsonia montana, specimens and roots; also a few roots of Thermopsis fraxinifolia. While digging one of these near the base of the mountain, struck upon a little clump of Schweinitzia, half buried in the leaves, five or six specimens; but a long hunt furnished no more.

Thursday, crossed Linville River in sight of the North Cove (Michaux's old residence) and went to Carson's on the Catawba. We lost a shoe from our black horse while descending the Blue Ridge, and he wore his hoof so as to lame him severely. Obliged to leave him at Carson's (as we could not exchange him to advantage) and hire another horse to take his place for a week. Crossed the Swananoa gap; got fine near view of Black Mountain; passed the night not far from its base (twelve miles from Asheville). Should have ascended, but could not do it so as to get back Saturday night to any place to stay, and longed to spend one Sunday in a civilized place where we could attend public worship. So went on to Asheville to dinner; passed Saturday afternoon in taking care of our plants. Heard very good preaching at the Methodist church on Sunday. Monday set out down the French Broad. Tuesday reached the Warm Springs; got a luxurious bath. Rode the afternoon through the rain to Paint Rock, etc.; stayed the night in Tennessee below. Got Buckleya in fruit, and other things I can't now specify. Wednesday, dug up Buckleyas, etc. Left Mr. Sullivant at Warm Springs, who, not being able to bear the absence from his wife and children longer, has left me alone with the team, and is by this time more than halfway to Columbus. Thursday, returned to Asheville. Friday, packed a fine box of roots, with which my wagon was loaded. Sent for my black horse. Saturday, bad weather; but made a little excursion on horseback, got roots of Arum quinatum, which, by the way, often has the lateral leaflets not at all incised, and then (in fruit) looks just like A. Virginicum. Buckley is often inquired after here, and seems to have been quite a favorite. He might have enlivened his journal had he informed us therein that he visited both Black and Bald Mountains with a merry company of ladies, and camped out on the summit! But the sly fellow kept all this to himself.

I begin to be in a hurry; but have yet much to do, and find it rather lonely. Monday and Tuesday I intend to devote to Hickory-Nut Gap, twenty-eight miles and back. Then visit Black, if I meddle with this mountain at all. Then, taking final leave of Asheville, go into the mountains near the head of French Broad, take up my quarters with a well-known hunter, try to reach Pilot and other high mountains which Buckley failed in reaching, and which have never been visited by a botanist, unless by Rugel;[134] thence to Table Rock, South Carolina, and by a roundabout way to Franklin, Macon County, Tolula Falls, and Clarksville, Georgia, where I shall try to sell out my horses and wagon, and take stage for Athens, where I am in the way to come by steam all the way to Princeton, via Augusta and Charleston, which bid fair to be healthy enough to warrant my passing through them without rashness.

It will be the 20th October ere I can hope to take you by the hand. Truly welcome are the newspapers you have kindly sent; but I hope for more by the next mail, for I have none later than the middle of September.

I never have been so hurried, and had so little time to write, but shall have the more to tell when I reach you, if it please Providence. Excuse chirography also, for pen and ink are wretched and my hands sore.

Aster Curtisii abounds and is very showy. A. Elliottii takes here the place of A. puniceus. I have found A. mirabilis.

Love to all, most warmly. Don't fail to mention me to dear Herbert.

Monday morning.—Off for Hickory-Nut Gap, where the scenery is said to be very grand, and the botanizing good. I am to get there Asplenium pinnatifidum, Stuartia pentagyna, and Parnassia asarifolia. Hard work, yet pleasant with a companion. I wish you could be with me.

Very pleasant Sunday service in the Presbyterian church here.

TO GEORGE ENGELMANN.

CAMBRIDGE, November 4, 1843.

I have been absent in the mountains of Virginia and Carolina—after live plants—from 11th August to yesterday; which will be my excuse for not replying to your letter of September 15th. I hope in the mean time you have found some way to send the roots you proposed. There are now connected express lines all the way through. L. & P. Franciscus & Company, No. 90 North Main St., St. Louis, are the agents of Brown & Company Express, Philadelphia; this connects with Harnden's Express to Boston, the speediest and cheapest method of sending when the package or box is not large, and speed is desirable....

Gaura Lindheimeri is a very fine plant, and flowered fully three months in our Garden. I am having a drawing—hoping to publish it sometime. I want more seeds of Œnothera rhombipetala. Ours flowered while I was away, and was killed by the frost, so that I secured no drawing. Send me all the seeds you can.

Inquire about the express to the East. We must somehow have the means of a more speedy and regular communication of parcels.

I found what I believe is your Lepidanche adpressa at Harper's Ferry, Virginia. Also some others in the mountains, which, with a few other plants, I will send to you by express soon....

You know I am obliged now to prepare for a terrible course of public lectures, to commence in February, so that I cannot work at the "Flora" until spring. But I will find time to study and revise any sets of Lindheimer's, Geyer's, and Lüder's plants you send....

As to my paper on Ceratophyllaceæ, I have long since wished it unpublished, as it contains mistaken views. So I do not care to distribute it.

February 2, 1844.

I have saved Gaura Lindheimeri by cuttings put in pots last autumn. We shall have it in flower early in the spring, and then shall exhibit it at the Horticultural Society's rooms in Boston.

TO HIS FATHER.

CAMBRIDGE, November 18, 1843.

MY DEAR FATHER,—The return of my birthday brings to mind, among other shortcomings, that I have neglected to write home since my return. I have been very busy, of course, since the 3d of the month, when I reached Cambridge, in answering the heap of letters that had accumulated, and in other business. And I have but just found time to commence the preparation of my course of lectures before the Lowell Institute, which is to commence on the 27th of February, and which will give me plenty of labor and anxiety until they are over....

I have laid in a good stock of health and strength for the labors of the winter—which I am like to need, for I have a great deal to do. Another year, if our lives are spared, I trust you will make me a visit here. I have just given notice that I shall wish to take possession of the Botanic Garden house (now rented to one of the professors) next autumn, where, if I can get a room or two furnished, I shall have a place to entertain you. Affectionate regards to mother and all the family.

TO JOHN TORREY.

CAMBRIDGE, February 17.

My time of trial draws near. A week from Tuesday I begin. There has been a pretty brisk application for tickets. But I have yet very much to do. My two last lectures are not even blocked out upon paper. Many pictures are yet to be made, and I shall have a busy time indeed until they are all delivered. The end will be deliverance indeed. Yet strange as it may seem, my spirits are rather on the rise; though I will not answer for them for ten days longer.

I have written an introductory which, with a few more touches, I shall be satisfied with. And some of my lectures which have least illustrations—such as that on food and nutrition—are pretty carefully written out. I have contrived a diagram illustrating the cycle of relations of three kingdoms, which I think is capital (as it is quite original), and which I long to show you. If I had three mouths more, I am convinced I could put my materials into the form of a capital course of lectures.

Zuccarini wrote me a year ago—when he sent the Japanese plants that we looked over together—that the Japanese species utterly confounded the difference between Rhododendron and Azalea; decandrous species having deciduous leaves, etc. If they must come together (and De Candolle seems doubtful) it would be a pity you did not follow that plan, as you early adopted it.

Then after all, in such case, are the Azaleas, as they will ever be called in cultivation, to make the section Azalea, or is A. procumbens to take that name?...

I wish you could see my Lowell anatomical illustrations. The pity is, that I shall hardly use them in this course, now that my introductory lecture only brings me down to them. (but I shall have them spread to look at), and I can only give to the subject about twenty minutes of my second lecture.

But it is very late indeed. Adieu.

Yours cordially,
A. GRAY.

March 1, 1844.

Well, you have heard what I had to say about my introductory lecture. I was satisfied. I said plainly what I intended to say and delivered it not very well indeed, but well enough to satisfy me that I could do well with practice. This evening I have made a second trial, and a more trying one by far. I have a cold and am a little hoarse, which was a good thing, for as to voice I filled the house. As I was full of illustrations, quite as much as would cover the whole side of a barn, I determined to try the experiment of lecturing by the general guidance of my notes only (which indeed were but partly written out). So with the long pole in hand to point at the pictures I set at work, and talked away for an hour and ten minutes.

I felt like a person who can hardly swim, thrown into the river, fairly in for it, and had to kick and strike to keep my head above water. The results are these. I was by no means satisfied, and thought I had made almost a failure. I left out many important points, I repeated myself a little now and then, and,—the usual result of extemporizing,—I did not get through, but was obliged to break off in the midst of the best of it. But, in spite of some difficulties of expression, and bad sentences, the whole was probably more spirited in appearance than if I had followed my notes. And the audience generally seemed more moved by it than by the first.

I consider it thus far successful; that under unfavorable circumstances, for I had no time to look over my notes beforehand, I made a desperate lunge, and yet avoided a real failure. It will place me so much at ease that I can hereafter, with or without notes, look fairly at my audience without wincing. So I shall do better hereafter....

I send you my notes (on Vacciniums) as far as written before I left for the South last summer; and with all Boott's memoranda as material. It would be crazy for me now to attempt to make any memoranda, or even to make the corrections that the new data require. Conclusions formed in hurly-burly are good for nothing; and I cannot, and must not, think of anything but my task. The two last of my lectures are not even arranged yet.

TO J. D. HOOKER.

CAMBRIDGE, 1st March, 1844.

MY DEAR FRIEND,—I was very much gratified at receiving your kind letter of January 16; and I was quite startled at the lapse of time, I assure you, when you reminded me that five years had elapsed since we were running about the streets of London together. Since that time you have seen the world, indeed, or some very out-of-the-way parts of it; and you now stand in a perfectly unrivaled position as a botanist, as to advantages, etc., with the finest collections and libraries of the world within your reach; and if you do not accomplish something worth the while, you ought not to bear the name of Hooker.

I thank you most cordially for all the news you kindly give me respecting the family, and wish to return my best thanks for being remembered to one and all. Your good old grandfather holds out so well that really I sometimes think I may yet take him again by the hand; for I long to make another visit to England. Perhaps I may in two or three years. But I hope ere that to see you here, where you may depend upon a most hearty reception; and the Greenes (who send remembrances) join me strenuously in begging you will make us a visit. After Sir William and Lady Hooker (seniores priores), whom we cannot expect to see under present circumstances, there is nobody in England I could so much wish to see as yourself.

Had I time, I should fill this sheet with gossip about my occupations, plans, and prospects. Of these hereafter, for I hope our correspondence will not end here. But I am now exceedingly pressed for time, having just commenced my course of public lectures in Boston on physiological botany. Indeed I have the second lecture to give this evening, and much preparation yet to make for it. But I must tell you that in August next I am to take possession of the house which belongs to our little Botanic Garden,—a quiet pleasant place, where I am to set up a bachelor establishment, have room enough for my herbarium, which I shall arrange à la Hooker, and a bed and a plate for a friend. So, if you wish to take an autumnal excursion, step on board the steamer and so drop in upon me some morning, where you may depend upon—in a humble way—as cordial a reception as I once received in Scotland.

Sullivant, who is a good, spirited fellow, is delighted at the thought of receiving a set of your cryptogamic collections. As to your generous proposal to send another to some public collection in this country, we will see. I will write something about it in due time.

TO JOHN TORREY.

CAMBRIDGE, 25th March, [1844].

I think I should be an unhappy, discontented, unthankful person not to be gratified with the success of my lectures. But it is not likely to turn my head. Everything proceeds quietly and soberly. I purposely directed no tickets to be sent to a paper that often reports lectures, as I did not wish it done. There has not been a line in the papers about the matter, except the very considerate notice about the beginning, which I sent you. My last week's lectures are called much the best. The first, on the anatomy and physiology of leaves, and exhalation and its consequences, occupied an hour and twenty minutes. My last, on food of plants, vegetable digestion, and the relations of plants to mineral and animal kingdoms,—in which I did my very best, and which required and secured the most intense attention on the part of the audience for a hundred minutes,—was received with an intelligent enthusiasm which did the audience credit. For it would be mere affectation for me to pretend not to know—as I well do—that it is one of the best scientific lectures that have ever been delivered in Boston. I have none left to compare with it. I have only four more to give, during which I dare say the interest will fall off; which will not disappoint or mortify me. From your truly kind remarks and warnings I suppose you look upon my success in this undertaking as extremely hazardous to my best interests. Now this duty came to me unsolicited and unexpected. I accepted it because I thought it was my duty to do so. Then I was of course bound to make every consistent effort to insure success. While viewing it at a distance, I felt much anxiety. But before I commenced, this entirely disappeared, and I have gone on just as coolly as you might do with your chemical course. I am thankful that (owing chiefly to the nature and novelty of the subject) I have done my work creditably. The little éclat which attends it, I am not so foolish as to care anything for, pro or con. It is entirely ephemeral. It may gratify my friends; but it does me no good, and I trust no harm. The general result may benefit the science of this part of the country. It will probably tend to advance my interests, as I certainly wish it may, the object of my ambition being high and honorable, as well as moderate....

Though I feel that I often—always—fail to do my whole duty, yet I do not feel, nor believe, that a perfectly consistent Christian course would expose me to persecution; nor that obloquy is a test of Christian character. These are to be borne like other evils, when they are incurred in the course of one's duty; but surely they are not to be sought, nor viewed as a test. Under the circumstances under which we are placed, would our unexpectedly meeting with obloquy be any test to us that we were doing right? Would it not lead us to suspect we had been at least unwise? Such men as Payson or Edwards, though they may often have been pitied, I suspect, were never persecuted. But, while I think you take a one-sided view and assume, an unscriptural test, in your own case, I thank you most sincerely for your kind admonition to me, and will try to profit by it. My sheet is fairly full.

I need not say how delighted I should be to see you here; but you must not come till the spring has fairly commenced, at least. The weather is excessively unpleasant, the roads almost impassable; it snows every three or four days, and not a speck of green is yet to be seen. A month later it will be comfortable here. I fear I shall not have a place to receive you before autumn, as a house is yet to be built for Dr. Walker. But I should still like to have a visit from you in the course of the summer.

Dr. Gray was always deeply interested in the religious thought of the day; reticent in regard to his own religious feelings and sensitive about any exhibition of them, he was ready at any time to discuss problems of theology and ecclesiasticism. His temper was naturally conservative, and he held by the habits of thought which had been early formed; but he was open to conviction, and by the process of his own thought broke through narrow bounds and rejoiced in all true progress in religion, both for himself and others. In the matter of scriptural authority, for example, he was in accord with Soame Jenyns, taking the ground quoted here:

"The Scriptures," says that writer, in his "Internal Evidences of Christianity," "are not revelations from God, but the history of them. The revelations themselves are derived from God, but the history of them is the production of man. If the records of this revelation are supposed to be the revelation itself, the least defect discovered in them must be fatal to the whole. What has led many to overlook this distinction is that common phrase that the Scriptures are the Word of God; and in one sense they certainly are; that is, they are the sacred repository of all the revelations, dispensations, promises, and precepts which God has vouchsafed to communicate to mankind; but by this expression we are not to understand that every part of this voluminous collection of historical, poetical, prophetical, theological, and moral writing which we call the Bible was dictated by the immediate influence of Divine inspiration."

He held this ground strongly when the general view of the Bible was narrower than of late years. As the years went on he grew broader and sweeter, feeling wider sympathy with all true, devout religious belief.

He was a constant church-goer, everywhere. When traveling he always made Sunday a resting-day if possible, and would go quietly off in the morning to find some place of service, in English if he could. He enjoyed the Episcopal service, though early habit and training had made him a Presbyterian; but, as he wrote in an early letter, "In fact I have no more fondness for high Calvinistic theology than for German neology.... But I have no penchant for melancholy, sober as I sometimes look, but turn always, like the leaves, my face to the sun."

He was a teacher in Sunday-schools in New York (the lady with whom he boarded has still a lively remembrance of his enthusiastic study of German that he might teach his class of German boys better), and also in his early years of Cambridge life, until the heavy load of work he was carrying made the Sunday more imperative as a day of rest. It was his rule to rest on Sunday. Rest for him was change of intellectual occupation, and he read all of the day he was not out at church; more especially on the philosophical questions, whether general or scientific, which next to botany were his chief interest. Books on these subjects were the few he bought outside of works on botany; as he said, he could only afford botanical books and had no money or room for general literature. He read the leading magazines, and occasionally biographies and travels, and if he had friends staying with him, Sunday was the day for talk and discussion. A friend writes such a lively reminiscence of one of these Sunday discussions, on a stormy winter day which shut all in the house, that it seems worth giving as a vivid description of him.

"Dr. Gray is more associated with the study and the room next it, but I recall him there (in the parlor) also, especially in the visit of which you wrote, made when Mr. John Carey was with you. He and the doctor held one Sunday a long discussion on the Ten Commandments as binding upon Christians. Mr. Carey argued that their only claim upon our obedience consisted in their having been re-ordained (indorsed as it were) by the church,—whether that meant the Holy Catholic or simply the Anglican Church was not decided, as I remember. Dr. Gray combated this extreme church view warmly and cleverly. Both were pugnacious amiably, as in their botanical fights. Both were excited, and the doctor showed his excitement in his characteristically self-forgetful way, by moving or jumping nervously about the room, sitting on the floor, lying down flat, but laughing and sending sparks out of his eyes, and plying his arguments and making his witty thrusts all the while. I enjoyed it very much, scarcely observing the odd positions any more than the doctor did. I had seen him so conduct himself before."

It may be added to this that Dr. Gray was noticeable throughout his life for his alertness. In the street he was usually on a half run, for he never allowed himself quite time enough to reach his destination leisurely. When traveling by coach and climbing a hill he would sometimes alarm his fellow-travellers by suddenly disappearing through a window in his eagerness to secure some plant he had spied; his haste would not suffer him to open a door. As his motions were quick, so that he seemed always ready for a spring, so he found instant relaxation by throwing himself flat on the floor when tired, to rest, like a child.

His physical characteristics expressed something of his mental qualities. He was quick and impetuous in temper, but his excitement was short-lived, and his prevailing spirit was one of apparently inexhaustible good-nature. He was the cheeriest of household companions; rarely was he depressed, only indeed when greatly fagged with some tremendous pressure of work or some worrying trouble difficult to settle; he was exceedingly hopeful, and always carried with him a happy assurance that everything was going on well in his absence; withal, he was fearless in all adventure, never willing to allow there had been any danger when it had passed! He was fond of arguing, but no partisan, so that however earnest and dogmatic be might seem, the moment the discussion was over there was no trace of bitterness or vexation left. He was a clear and close reasoner himself, and thus impatient of defective reasoning or a confused statement in others. He was quick, too, in turning his opponents' weapons against them; sometimes he would escape from a dilemma in a merry, plausible form, but in serious argument he always insisted upon downright sincerity.

TO W. J. HOOKER.

April 1, 1844.

I finish my course of Lowell lectures this week, which have succeeded beyond my most sanguine expectations. I have restricted myself to physiological botany only,—taken up only great leading views,—used very large paintings for illustrations, six to eight feet high, which the great size of the room required, and then have given to sound scientific views a general popular interest.

TO JOHN TORREY.

CAMBRIDGE, May 24, 1844.

I have been using Dr. Wyman's microscope of late, and it works well. By the way, I have been studying fertilization a little, and have got out pollen-tubes of great length; have followed them down the style, have seen them in the cavity of the ovary, and close to the orifice of the ovule.

My first views were in Asarum Canadense and A. arifolium, where I can very well see the pollen-tubes with even my three-line doublet! I have seen them finely in Menyanthes; and in the ovary in Chelidonium!

I am lecturing[135] in a popular and general way entirely on physiological botany, and offering no encouragement to any to pursue systematic botany this year. My great point is to make physiological botany appear as it should be,—the principal branch in general education. Next year I hope to take up the other branch.

I am using the Lowell illustrations (though too large for my room), and am having no additional ones made for the college. For simple things I depend much upon the blackboard. I have given two lectures on the longevity of trees, and have a third yet to give, or at least half of another....

The plants from the mountains have some done well, others poorly. Buckleyas had a hard time of it. Many are dead; none I think will flower this season, as they only put out from the root. Diphylleia, Saxifraga Careyana, a new one like it, also S. erosa, etc., are now in flower. Astilbe is in bud, also Vaccinium ursinum. One Carex Fraseri flowered. Hamiltonia only starts from the root.

In 1844, finding he needed more room for his rapidly increasing herbarium, Dr. Gray applied for the use of the Botanic Garden house, which since the death of Dr. Peck had been occupied for a while as a boarding-house, and later by Dr. and Mrs. Walker. He moved into it in September, and there remained until the end of his life. He had a great attachment for the house, as the only one in which he had resided for any length of time; and it saw the gradual growth of his herbarium, needing before many years the addition of a wing to give more room, until, having overrun all possible places for its accommodation, it was removed in 1864 into the fireproof building which now holds it.

The garden was laid out by Dr. Peck in 1808, and the house built for him was finished in 1810. Mr. Nuttall, the botanist and ornithologist, who boarded in it while giving instruction in botany, left some curious traces behind him. He was very shy of intercourse with his fellows, and having for his study the southeast room, and the one above for his bedroom, put in a trap-door in the floor of an upper connecting closet, and so by a ladder could pass between his rooms without the chance of being met in the passage or on the stairs. A flap hinged and buttoned in the door between the lower closet and the kitchen allowed his meals to be set in on a tray without the chance of his being seen. A window he cut down into an outer door, and with a small gate in the board fence surrounding the garden, of which he alone had the key, he could pass in and out safe from encountering any human being.

The garden, though small, was planned with much skill, and when Dr. Gray first lived on the place was much more filled up in the centre with trees and shrubs, so that since one was unable to see from one path to another, it seemed much larger than when more open. Dr. Peck, who had visited Europe and learned much of botanical gardens there, when complimented on his success in laying it out, said that "he felt he had been at work on a pocket-handkerchief!" Dr. Gray, as his letters show, fell earnestly at work to restock the garden, and from his various journeys, his correspondents, and the many seeds and roots which were coming in from the Western explorations soon made it a valuable spot for exchange. It is interesting to note how many plants, now the common stock of all gardens, were first grown and flowered here. One bed for many years always went by the familiar name of "Texas," as being the place where the new Texan seeds were grown. The fund for endowment was very small, and added greatly to the care of its oversight, because of the effort to keep within the income. For two years after Dr. Gray was living in the Garden house, he gave up two bed-rooms to the greenhouse plants, and so saved the Garden the expense of fuel for that period! One of his first deeds was to abolish the fee and make admission to the Garden free. It was the first—and remained for more than sixty years the only—public botanic garden in the country.

TO JOHN TORREY.

Tuesday evening, October 1, 1844.

I am about half fixed at the Garden, and shall probably sleep there to-morrow night. Were it not that my woman-kind has disappointed me, we should dine there to-morrow....

Dr. Wyman[136] wishes much to accompany Frémont if he goes on another journey, entirely at his own expense, if need be. As his object is entirely zoölogy, he will not interfere with Frémont's botanical plans, while the results would redound to Frémont's advantage. He is a most amiable, quiet, and truly gentlemanly fellow, retiring to a fault, but full of nerve, and surely is to be the great man of this country in the highest branches of zoölogy and comparative anatomy. I therefore very strenuously solicit your influence at court in his behalf.

I am glad that Frémont takes so much personal interest in his botanical collections. He will do all the more. I should like to see his plants, especially the Compositæ and Rosaceæ. As to Coniferæ he should have the Taxodium sempervirens, so imperfectly known, and probably a new genus. Look quick at it, for it is probably in Coulter's collection which Harvey is working at....

Cordially yours,
A. GRAY.

February 12, 1845.

My first lecture is to-day finished, and has this evening been read to Mr. Albro.[137] Half of it is devoted to a serving up of "Vestiges of Creation" (which Boott says is written by Sir Richard Vivian), showing that the objectionable conclusions rest upon gratuitous and unwarranted inferences from established or probable facts. Peirce is examining Mulder,[138] that we may fairly get at his point of view. His conclusions as to equivocal generation are non-constat from his own premises. On the whole series of subjects Peirce—who is much pleased with the way I have put the case in my introductory—and myself think of concocting a joint article, though my time will prevent me from working out some of the subsidiary points just now.

I assure you I am quite well and hearty, just in capital working mood. As to the lectures, I must work hard all the way through, but do not feel any misgivings. My house is hot enough, I assure you; no trouble on that score. As to spontaneous generation, the experiment of Schultz[139] is nearly or quite a test, and goes against it. Love to all.

Ever yours,
A. GRAY.

The next letter contains the first allusion to Isaac Sprague, so long associated with Dr. Gray as illustrator of his works. Isaac Sprague was born in Hingham in 1811. He early showed a faculty for observation, and a gift for painting birds and flowers from nature. His talent was discovered, and he was invited by Audubon in 1843 to join his expedition to Missouri, and to assist in making drawings and sketches. President, then Professor, Felton, having met him in Hingham, and knowing Dr. Gray was looking for some one for his scientific drawings, recommended Mr. Sprague, and he began with the illustrations for the Lowell lectures and the new edition of the "Botanical Text-Book." Dr. Gray was delighted with his gift for beauty, his accuracy, his quick appreciation of structure and his skill in making dissections. Mr. Sprague was from that time the chief, and mostly only, illustrator for his books, both educational and purely scientific.

Dr. Gray is said to have stated that Mr. Sprague had but one rival,—Riocreux; and he considered that draughtsman's classical drawings inferior to Mr. Sprague's.

TO JOHN TORREY.

CAMBRIDGE, March 8, [1845?]

... I finish Lichens this afternoon; and have next two lectures on Fungi and spontaneous generation to give. I interweave a good deal of matter, such as, on Ferns, the part they played in the early times of the world, à la Brongniart. Mosses, filling up lakes and pools; Sphagnum, Peat. Lichens, first agents in clothing rocks with soil. I have noble illustrations of rust in wheat, ergot, etc., and Sprague is now hard at work on smut, à la Bauer.

You remember the letter I sent you from Prestele of "Ebenezer, near Buffalo," and which you still hold. Well, he has sent me for inspection a most superb set of drawings, both of cultivated and of some native plants, exceedingly well done. Also specimens of his work in cutting on stone, which he does admirably. He did the work in Bischoff's "Terminology," which perhaps you remember, two quarto volumes. What a pity he did not have the State-Flora plates to execute!

If Dr. Beck and yourself go on with your plan, he is your man to engrave the plates on stone. Our Illicium is now in full flower; but I cannot spare Sprague a moment to draw it yet; unless, indeed, it is quite certain you will want it this year, when I would try. He must work hard for me two weeks longer....

My cutting up of "Vestiges of Creation" was a fine blow, and told. Peirce, who you know was rather inclined to favor Rogers a while ago, is now sound and strong. We think of sending a critical analysis of the first part of Mulder, as our joint work (if he finds time to put in form the physiological deductions I give him), to the meeting of geologists and naturalists at New Haven next month.

Mulder is very ingenious; but we can blow up the whole line of his arguments, and show that it all amounts to nothing; that he has not in this advanced our knowledge a particle; and that his generalizations are unsound. Why did you not have a part of my article reprinted in New York? That would be the best reply to all his stuff.

The printing of my book will be through next week.

March 30.

I am now half through, and have got almost done with Fungi. The audience take so much to the "Cryptogamic matters," especially the afternoon audience, which is as a whole the most intelligent and refined, that I let them run on, and they will occupy the whole course, except three lectures. I gave one lecture, generally thought nearly the best, on the large Fungi, mushrooms, truffles, morels, puff-balls, with some good general matters. To-day I have taken the small ones, moulds, mildews, rust, and smut in wheat, with superb illustrations. Ergot is still left over, along with the diseases in potatoes, the plant of fermentation, the Botrytis that kills silk-worms, with some recapitulatory matters on spontaneous generation, which must be cooked up for Friday. Then comes Algæ; the large proper ones (Lecture 8), of which a fine series of illustrations is now nearly done.

Lecture 9. Then the low, minute forms and Confervæ come, and gory dew, red snow, superbly illustrated, ending with diatoms, transitions to corallines through sponge, etc., and the locomotive spores of Confervæ, Zoösporæ.

Lecture 10. Whole subject of spontaneous movements and sensibility in flowering plants, the life of plants, etc. (treated in a somewhat original way), and the real differences between plants and animals.

Lecture 11. The principles of classification. Individuals, species, their permanence, genera, orders, etc.

Lecture 12. Historical development. The Linnæan system, the natural. This ends so as to give me a fine place to begin at next year....

I shall soon be able to spare Sprague to draw the Illicium, if it still holds on. But I cannot spare him just yet. He has still to copy the red-snow bank from Ross, eighteen feet long!—finish two pieces of algæ, etc., etc.

TO A. DE CANDOLLE.

April 5, 1845.

I anxiously wait for the notices of the life and writings of your lamented father, which you so kindly offer. I agree with you that that of Daubeny[140] gives the best view of the philosophy of his science; and yet there are points of view that he has not touched upon. You, of course, know better than any one else what were your father's philosophical views in natural history, his modes of thinking and working; and if, when you send me the above-mentioned documents, you would also feel at liberty to place such confidence in me as to give me your own views and suggestions upon the subject, and especially upon the points that other writers appear to have overlooked, I should be able to produce, in the "North American Review," a much more important article and a worthier tribute to the memory of one so revered on this side of the Atlantic as well as in Europe. May I hope you will favor me in this respect?

Many thanks for the botanical news. I long to be delivered from the pressure of the engagements that have consumed so much of my time for the last year or two, and finish the "Flora of North America."

I remain, ever, my dear friend, faithfully yours,

A. GRAY.

TO JOHN TORREY.

August, 1845.

The new post-office law is an excellent thing, as it enables us to exchange our missives frequently, to send little pieces of news, and ask and answer questions without waiting for time and matter to fill up a formal letter.

I must tell you a little change made in my sanctum here. You are to imagine me writing at a sort of bureau-escritoire (standing under Robert Brown's picture), which I fortunately picked up the other day for $10. It is of old dark wood a century old, and contains below four drawers, while the upper part, which opens into a fine writing-table, has eight pigeon-holes, six drawers, and a little special lock-up with several drawers and pigeon-holes more. You know I like any quantity of these stowaway places. I have sent upstairs the table which stood in its place, and brought down the round one, so that I have more room than before.

TO W. J. HOOKER.

October 14, 1845.

Your excellent father lived to a truly patriarchal age. Mine, who has been in failing health for some time, I learn to-day is suddenly and extremely sick, and I set out for my birthplace immediately, in hopes yet to see him once more.

His father died October 13, before he reached Sauquoit. He had made his son a visit in Cambridge after he was established at the Garden house, more especially to consult a physician for his failing health.

TO JOHN TORREY.

CAMBRIDGE, November 15, 1845.

My visit to Oakes[141] I was chiefly to this intent. You know that I have been waiting and waiting for Oakes to give, not his New England "Flora" (which I fear he will always leave unfinished), but a predromus of it, for my use and for New England. The consequence of waiting is that Wood[142] is just taking the market, against my "Botanical Text-Book," mostly by means of his "Flora." Letters from Hitchcock—and elsewhere—all point to the probability that they will have to use his book (of which, by the way, he is preparing a second edition, which he cannot but improve), and ask me to prevent it by appending a brief description of New England or Northern plants to my "Botanical Text-Book." A plan has occurred to me by which this might be done, were it not that I will not tread on the heels of anything that Oakes (who has devoted a life of labor to this end) will actually do.

As something must be done at once, I have proposed to Oakes to make myself the necessary conspectuses of orders, analyses, etc.; to join the proposed thing on, or to dove-tail it into, the "Text-Book;" and also to furnish the generic characters, and he is to write the specific characters and all that for New England plants. I give him as limit 250 pages brevier type, 12mo (say 300), and insist upon having the greater part of the copy on the 1st March, and that it shall be published on the 1st April. That I may cover the ground of Wood, and introduce it into New York, I propose, if you think it right and proper, to add the characters of the (about 150) New York plants not found in New England, distinguishing that by a †.

Oakes promises to do it. But our understanding is explicit that if he cannot get through with it in time, he is soon to let me know, and to furnish me with New England matters, when I am to do, not exactly this, but a more compendious manual of the botany of New England, New York, New Jersey, and Pennsylvania, that is, the Northern States proper. It will be imperfect and hasty, but it will prevent Wood from fixing himself so that he cannot be driven out.

I propose to have a sufficient number of copies of this (in whatever form it may appear) bound up with the "Botanical Text-Book" to meet the demands of the one-book system in New England and New York, and to afford it at a price reduced to a minimum, so that nothing is to be made out of it, at least out of the first edition.

How does this all strike you? I am convinced that something must be done, and I will see if we can't have a very popular, and at the same time a pretty good book.

George[143] sends his warm regards.

21st November, 1845.

I have driven Oakes so absolutely into a corner that I think he will work for once. The man's preparations and materials are enormous! and for his sake I hope he will. If he does not, I shall know in time,—that is, as soon as I can use the knowledge,—and then the plan may take such form as may be deemed best. I should then wish to make it more absolutely a supplement of "Botanical Text-Book;" but only for the proper North. In the way in which it would then be done, with Persoonish[144] compactness and brevity, I doubt if you would care to engage in it. As soon as we can get out the proper Botany of the United States, I should wish it to supersede this to a great extent. In my hands, I would sell it so cheaply as to make very little, except as it promotes the sale of the "Botanical Text-Book." I would sell the "Text-Book" with it for $2, or less even. The great object is to keep the ground clear by running an uncompromising opposition against the threatening interlopers.

My lectures are to commence January 13th.

TO J. D. HOOKER.

CAMBRIDGE, 31st December, 1845.

I was much pleased to receive your pleasant letter of the 29th October last, and I read with interest the account of the debate on the occasion of the election by the Edinburgh Town Council. Such defeats can do you no harm. I suppose you are now going on with the "Flora Antarctica." I need not say that I should be very glad to see the Antarctic plants of the Wilkes Expedition in your hands. The botanist who accompanied the expedition is no doubt perfectly incompetent to the task, so greatly so that probably he has but a remote idea how incompetent he is. I have not seen him nor the plants. Certainly I would not touch them (any but the Oregon and Californian) if they were offered to me, which they are not likely to be. I consider myself totally incompetent to do such a work without making it a special study for some years, and going abroad to study the collections accumulated in Europe. Of course if they are worked up at all in this country, they will be done disgracefully. I publicly expressed my opinion on the subject in "Silliman's Journal." But I have long been convinced that nothing can be done. The whole business has been in the hands till now of Senator——, the most obstinate, wrong-headed, narrow-minded, impracticable ignoramus that could well be found.... If to this you add an utter ignorance of those principles of comity and the spirit of interchange that prevail among naturalists, and a total want of comprehension of what is to be done in the scientific works in question, and you will see that nothing is to be expected from such sources. They have thrown every obstacle they could in the way of their naturalists—Dana and Pickering, for instance—so much so that Pickering, though a patient man, once threw up his position in disgust, I have heard, but, by some concessions made to him, was finally persuaded to retain it.

Some of the scientific reports will soon be published, Dana on the Corals, etc., which will, I suppose, be very creditable to him. When any of the volumes appear I am somewhat inclined to call public attention to some of this gross mismanagement and incompetency in these wrong-headed managers, in a review. I thank you very much for all the botanical news you give, and hope you will still favor me now and then with other such epistles.

I have never worked so hard as for the last four years, nor accomplished so much. Still it will not show for much in your eyes, and I receive many an exhortation like yours to go on with the "Flora." But a world of work that could only be done by myself, the pressure of the duties of my new position, and the necessity of taking, indeed of creating, and maintaining a stand that should make my department felt and appreciated, has indeed sadly interrupted the work which I am of all others most desirous to complete. I have already a great deal of matter in a state of forwardness, and another year (Deo favente) will, I trust, give you a better account of me. My last course of public lectures in Boston commences in a fortnight, and will be over towards the close of February. You will admit that there is some temptation to a person who has so many uses for money, when I tell you that I received twelve hundred dollars for the delivery of twelve lectures, and that there are strong reasons beyond what the institution that employs me may justly demand, that I should do my best. This, however, will soon be over, and the "Flora" shall be pushed with vigor.... I greatly long to revisit England and to see you all once more. Nothing would delight me more; and there is a world of work I want to do in the collections of England and the Continent. Indeed you may look to see me one of these days, for I cannot long be satisfied or quiet without such a visit; though I shall hardly dare to show my face till the "Flora" is finished. How glad I shall be to see you in your quarters at Kew, and renew my acquaintance with all the family, of whom I retain so many pleasant memories. With kind regards to all, believe me,

Ever your affectionate friend, A. GRAY.

TO JOHN TORREY.

CAMBRIDGE, January 26, 1846.

Your favor of the 22d I found this evening on my return from my afternoon's lecture. I am very tired and cannot write much this evening. Four of my lectures[145] are off. You will be glad to know that they have gone off very well—the three first admirably; indeed I was surprised myself at the fluency, ease, and "enlargement" which was given me. The fourth, both last evening and this afternoon, was poorer—interesting details, but scrappy, and less comfort in speaking. Splendid illustrations up though.... The pictures were worth something, if the lecture was not. I shall spur myself up hard for those four to come, which are fully illustrated, in fact a complete *embarras de richesses*. Then come the four geographical lectures, which if Sprague gets the illustrations ready will be very interesting, I think. I must work them off well, for at least two of our seven members of corporation are constant hearers.

... There is a formidable amount of work of various sorts that should be accomplished (Deo favente) before the July vacation.... The contemplated expedition is a land one, from Lake Superior by North Pass to upper Oregon, down to Lewis River; up that, and then over to the Gila River in California. I know of no botanist to go. Can you find one? Sprague cannot be spared, and will not leave his wife and family for so long.

... Some of our Congressmen must feel a little ashamed that England is so cool and quiet in spite of all their bluster. Capital for peace that the Peel ministry is still in. We owe much gratitude to the new Lord Grey....

TO GEORGE ENGELMANN.

CAMBRIDGE, April 8, 1846.

What is Lindheimer about? Why is not his last year's collection yet with you? We have just got things going, and we can sell fifty sets right off of his further collections, and he can go on and realize a handsome sum of money, if he will only work now! And he will connect his name forever with the Texan Flora!

I am at the "Flora" again and hope to do great things this year,—shall work hard and constantly.

Besides, by the aid of my young and excellent artist Sprague's drawings, and Prestele to engrave cheaply and neatly on stone, I am going to commence a Genera Illustrata of the United States, like T. Nees von Esenbeck's "Genera Germanica Iconibus Illustrata."—the plates to be equally good, and quite cheap too. The first volume, one hundred plates, going on regularly from Ranunculaceæ, will be preparing this summer, and will be out in the fall....

May 30.

Have done something at the "Flora;" shall do much work this season after July 4th, when college duties are over. Drawings for "Genera" are getting on well.

One word now on another point. We must have a collector for plants living and dry to go to Santa Fé, with the Government Expedition. If I were not so tied up, I would go myself. Have you not some good fellow you can send? We could probably get him attached somehow so as to have the protection of the army, and if need be I could raise here two hundred dollars as an outfit. He could make it worth the while. He could collect sixty sets of five hundred plants (besides seeds and Cacti) very soon, which, named by us, would go off at once at ten dollars per hundred. Somebody must go into this unexplored field! Let me know if you think anything can be done, and I will set to work. The great thing is a proper man.

July 15.

I duly received your favor of June 25th; am delighted that you found a man to send to Santa Fé. I approve your mode of carrying out the plan, and will not be slow to aid in it. I wrote at once to Sullivant, telling him to forward fifty dollars for Fendler,[146]—to take his pay in Mosses and

Hepaticæ, and to give instructions about collecting these, his great favorites. Before this reaches you, I am sure you will hear from him. He is a capital fellow, and Fendler must be taught to collect Mosses for him.

Then came your letter of July 3d. All right. I immediately wrote to Marcy, the Secretary of War, and to Colonel Abert, the head of the Topographical Engineer Corps; asked for protection and transportation; told the secretary to send anything he might be disposed to do to you at St. Louis. I then inclosed your letter to Mr. Lowell, and have just received it back again, with his letter, which I inclose to you! Is it not handsome?... Now Fendler has money enough to begin with. As soon as he is in the field, and shown by his first collections that he is deserving, I can get as much more money advanced for him, from other parties. If he only makes as good and handsome specimens as Lindheimer, all will be well. His collections should commence when he crosses the Arkansas; his first envoi should be the plants between that and Santa Fé, and be sent this fall, with seeds, cacti, and bulbs, the former of every kind he can get. These must be confined to yourself, Mr. Lowell, and me, till we see what we get by raising them. Other live plants he had better not attempt now.

His next collection must be at and around Santa Fé. But instruct him to get into high mountains, or as high as he can find, whenever he can. The mountains to the north of Santa Fé often rise to the snowline, and are perfectly full of new things. But you can best judge what instructions to give him. We can sell just as many sets of plants as he will make good specimens of. But forty sets is about as many as he ought to make....

It is said that a corps of troops is to be sent up through Texas towards New Spain. Lindheimer ought to go along, and so get high up into the country, where so much is new, and the plants have really "no Latin names."

October 8th.

By the way, meeting Agassiz last evening, I was pleased to learn that he claimed you as a schoolmate, and spoke of you with lively pleasure. He is a fine, pleasant fellow. We shall take good care of him here.

January 5, 1847.

I am glad so fine a collection is on the way from Lindheimer, and greatly approve his going to the mountains on the Guadaloupe. How high are the mountains? If good, real mountains, and he can get on to them, and into secluded valleys, he will do great things....

We will keep ahead of the Bonn people. By the close of next summer (Deo favente) we may hope to have the botany of Texas pretty well in our hands.

Do you hear from Fendler? Hooker says that region, the mountains especially, is the best ground to explore in North America! There is a high mountain right back of Santa Fé. Fendler must ravish it.

TO JOHN TORREY.

Wednesday, [October, 1846].

A Mr. Baird,[147] of Carlisle, Pa., called on me yesterday, evidently a most keen naturalist (ornithology principally), but a man of more than common grasp. He talked about an evergreen-leaved Vaccinium, which I have no doubt is V. brachycerum, Mx., that I have so long sought in vain!...

13th October, 1846.

I leave Agassiz in New York. He will leave New York Wednesday morning; join me at Princeton, and go on with us to Philadelphia that evening. We shall probably go together to Carlisle, where he has something to do with that capital naturalist, Professor Baird, and I have to get live Vaccinium brachycerum. He will soon return to make ready his lectures here.

Agassiz is an excellent fellow, and I know you will be glad to make his personal acquaintance. I must make my stay, such as it can be, at Princeton, on my return....

9th December, 1846.

Agassiz lectured first last evening; fine audience; he had a cold; was very hoarse, so that he spoke with discomfort to himself, but it went off very well. Though he by no means did himself justice, the audience seemed well pleased, and the persons I spoke with at the time, the most intelligent people, were quite delighted and impressed. He has repeated to-day. I expect to hear him again on Friday....

I have sixteen proofs of "Genera Illustrata." The engraving is clean and neat, but except a few of the last, they are not done so well as we expect, and do not do justice to the drawings, which, indeed, are almost matchless. Prestele has, in some, altered the arrangement of the analyses on the plate; consequently they must be done over again.

I am clear that Prestele can do what I want, so I have given him further instructions, and have raised his pay to $2.50 each; increasing my own risk thereby. Sprague has discovered some new quiddities about the position of the ovule in Ranunculaceæ. The raphe is dorsal in all of them, with pendulous ovules; also in Nelumbium.

He will go on very slowly; I can't hurry him. He has not yet taken up Croomia.

You have not told me about Chapman's queer plant yet!...

Unless Nuttall has arrived, which I do not hear of, it is too late for him till next fall; for his object was to secure three months' absence out of the present year, and three out of next.[148]

January 24, 1847.

Agassiz has finished his lectures with great eclât—most admirable course—and on Thursday evening last he volunteered an additional one in French, which was fine.

I gave you the explanation you asked for in my last letter, which I still hope you will find. What I then said about the excellent tone of his lectures generally was fully sustained to the last; they have been good lectures on natural theology. The whole spirit was vastly above that of any geological course I ever heard, his refutation of Lamarckian or "Vestiges" views most pointed and repeated. The whole course was planned on a very high ground, and his references to the Creator were so natural and unconstrained as to show that they were never brought in for effect.

The points that I. A. Smith has got hold of were a few words at the close of his lecture on the geographical distribution of animals, in which he applied the views he maintains (which are those of Schouw still further extended) to man.

He thinks that animals and plants were originally created in numbers, occupying considerable area, perhaps almost as large as they now occupy. I should mention that he opposes Lyell and others who maintain that very many of the Tertiary species are the same as those now existing. He believes there is not one such, but that there was an entirely new creation at the commencement of the historic era, which is all we want to harmonize geology with Genesis. Now, as to man he maintains distinctly that they are all one species. But he does not believe that the Negro and Malay races descended from the sons of Noah, but had a distinct origin. This, you will see, is merely an extension of his general view. We should not receive it, rejecting it on other than scientific grounds, of which he does not feel the force as we do.

But so far from bringing this against the Bible, he brings the Bible to sustain his views, thus appealing to its authority instead of endeavoring to overthrow it. He shows from it (conclusively) that all the sons of Noah

(Ham with the rest) were the fathers of the extant Caucasian races,—races which have remained nearly unaltered from the first, and that if any negroes proceeded from Ham's descendants, it must have been by a miracle. That is the upshot of the matter. We may reject his conclusions, but we cannot find fault with his spirit, and I shall be glad to know that Dr. I. A. Smith, in the whole course of his public teaching, has displayed a reverence for the Bible equal to that of Agassiz. I have been on the most intimate terms with him: I never heard him express an opinion or a word adverse to the claims of revealed religion. His admirable lectures on embryology contain the most original and fundamental confutation of materialism I ever heard.

I make the "Manual" keep clear of slavery,—New Jersey, Pennsylvania (if little Delaware manumits perhaps I can find a corner for it), Ohio, Indiana or not as the case may be, leave out Illinois, which has too many Mississippi plants, take in Michigan and Wisconsin, at least Lapham's[149] plants near the Lakes. That makes a very homogeneous florula.

I have made as usual much less progress than I supposed; so now, pressed at the same time with college duties, I have to work very hard indeed. Carey is coming on to help me.... Sheet full.

July 20, 1847.

Did you not know that an application has come from Wilkes through Pickering[150] to Sprague to make some botanical drawings for the Exploring Expedition, which, as I supposed they were to be for your use, I persuaded Sprague to promise to undertake, at ten dollars for each folio drawing with the dissections full.... The price we fixed is as low as Sprague can do them for, to any advantage, even if he had nothing else to do. The price I fixed for the drawings of "Genera," and which I thought very large, ($6 per plate) does not thus far pay Sprague day wages, he takes so much time and care with them. I can only hope that the experience and facility he is getting will enable him to knock them off faster hereafter. You see therefore that Sprague cannot afford to make the drawings for Emory at the price he made those for Frémont—two dollars apiece. He will do them better; having now such skill in dissections he will display structure finely, but he must not undertake them under six dollars apiece, since they will cost him as much time as do my octavo "Genera" drawings. He might make what you want along this summer and autumn; I am not crowding him.

September 28, 1847.

I had a pleasant visit to Litchfield of three days, including the Sabbath. On the banks of a lake in the neighborhood I stumbled on a species of Cyperus dentatus, which in the "Flora of the Northern States" you credit to Litchfield, Brace.[151] This Mr. Brace, who is an uncle of J.'s, I met for a moment at New Milford, where he now lives. There are three great aunts, most excellent old ladies, who live in a simple and most delightful manner at Litchfield. The youngest, who has been J.'s guardian almost from infancy, returned with us to Boston for a week or two. Their brother, Mr. Pierce, who died only last year, was, it seems, an old friend of yours, through whom they feel almost acquainted with you. He passed a part of his life in New York, was a mineralogist, and I think I have seen his name as a member of the Lyceum. Pray tell me about him.

I found it not easy to make an arrangement in New York for the publication of the "Illustrated Genera," by which I could get back directly the money I have expended in it. I think, therefore, I shall go on to defray the expenses of the first volume myself, which I think I shall be able to do, and thus manage to get the immediate proceeds myself. As to the "Manual," I have unwittingly made it so large, in spite of all my endeavors at compression, that I can make nothing to speak of from the first edition, even if it sells right off.

TO J. L. L.

Monday evening, 9 o'clock, 1847.

When I reached home Henry and Agassiz were here. No one else came (as I expected), and Agassiz insisted on returning in the nine o'clock omnibus. Agassiz and Henry enjoy and admire each other so richly, and talk science so glowingly and admiringly, that I think I should not have been at all surprised to see them exchange kisses before they were done. And Agassiz told him he meant to come to Cambridge, and they began to talk of their children, and Agassiz read extracts from letters just received from his wife and his son, who—to Agassiz's great pride and satisfaction—had just climbed the Fellenberg in the Breisgau, slept on the summit in the open air to see the sun rise in the morning, then descended and walked, I forget how many miles. Pretty well for a lad of eleven.

It is not a year since I told Henry that he should have either Agassiz or Wyman at Washington, but that we must have one of them at Cambridge. Beyond all expectation we have them both!

Henry gave me—I know not what led to it—a full detailed account of his life from early boyhood, which was full of curious interest and suggested much matter for reflection. In the evening we fell to discoursing on philosophical topics, and Henry threw out great and noble thoughts, and as we both fell to conversing with much animation my headache disappeared entirely. There is no man from whom I learn so much as Henry. He calls out your own powers, too, surprisingly....

I have been addling my brain and straining my eyes over a set of ignoble Pond-weeds (alias Potamogeton) trying to find the

"difference there should be
Twixt tweedle-dum and tweedle-dee,"

and wasting about as much brain in the operation as your dear paternal would expend in an intricate law case, for all of which I suppose nobody will thank me and I shall get no fee. Indeed, few would see the least sense in devoting so much time to a set of vile little weeds. But I could not slight them. The Creator seems to have bestowed as much pains on them, if we may use such a word, as upon more conspicuous things, so I do not see why I should not try to study them out. But I shall be glad when they are done, which I promise they shall be before I sleep.

10.45 P.M.—There, the pond-weeds are done fairly.

TO W. J. HOOKER.

CAMBRIDGE, December 1st, 1847.

I reply early to your kind letter of October 30th to assure you that I shall with much pleasure contribute so far as I have opportunity to the new Botanical Museum, which, under your charge, and with your great opportunities for obtaining things from every part of the world, will soon become a magnificent collection. I have already several things to send you, such as two very large entwined stems of Aristolochia Sipho, which I brought from the mountains of Carolina.; a Dasylirion from Texas, etc. I have some time ago made arrangements for getting curious stems from Para, through a friend in Salem, who will also incite the masters and supercargoes of ships from that port which trade with various out-of-the-way parts of the world. The first things sent from Para were slabs rather than truncheons of wood (all ordinary exogenes), but I am promised palm stems and woody climbers, of which I shall take a portion to build up our general Natural History Museum at Cambridge, which with the zeal of Agassiz and Wyman is now likely to grow; the rest I will send to you. If you will send me a few duplicates of your circular, I will have them placed in proper hands where they may turn to good account. I am delighted to hear such pleasant things of Dr. Hooker, which I had also heard last summer from Mrs. McGilvray. I owe him a letter, but it is too late to send my congratulations, now that he is probably far on the way to India. I admire his zeal and energy, and wish him an excellent time and a prosperous return. The government has behaved most handsomely in affording him such important aid in his undertaking.

Proper specimens of maple sugar will keep perfectly well if placed in a glass jar with a closed cover. I will surely send some in the spring.

TO GEORGE ENGELMANN.

CAMBRIDGE, December 20, 1847.

I got a parcel from New York on Saturday evening, containing a few welcome plants of Wislizenus'[152] collection, and a set of Fendler's from Santa Fé, up to Rosaceæ. The specimens are perfectly charming! so well made, so full and perfect. Better never were made. In a week I shall take them right up to study, and they are Rocky Mountain forms of vegetation entirely, so I can do it with ease and comfort. It is a cool region that, and dry. If these come from the plains, what will the mountains yield? Fendler must go back, or a new collector, now that order is restored there.

All Fendler's collection will sell at once, no fear, such fine specimens and so many good plants. Pity that F. did not know enough to leave out some of the common plants, except two or three specimens for us, and bestow the same labor on the new plants around him.

Send on the rest soon.

Yours cordially,
A. GRAY.

TO CHARLES WRIGHT.

CAMBRIDGE, January 17, 1848.

DEAR FRIEND,—That I ought to have replied to your letter of the 19th November, to say nothing of that of September 21 and June 18, there is no doubt. The letter I have carried in my pocket a good while, hoping to catch a moment somewhere and some time to write to you, especially as the time approaches in which I may be sending a parcel to New Orleans for you. But I have not had an hour's leisure not demanded by letters of immediate pressing consequence, or in which I was not too tired to write.

There are many correspondents whom I have neglected almost as much as I have you. I have worked like a dog, but my work laid out to be finished last July is not done yet.

But from about the time of your last letter a providential dispensation has prevented me from doing what I would, namely, the sickness, by typhoid fever, of a beloved brother (a Junior in college here), who required every leisure moment from the time he became seriously sick up to the 9th inst.—a week ago—when it pleased the Sovereign Disposer of events, to whom I bow, to remove him to a better world; and I am but recently returned from the mournful journey to convey to the paternal home (in western New York) his mortal remains. This has somewhat interrupted the printing of the last sheets of my "Manual of North American Botany;" which, with all my efforts at condensation, has extended to almost eight hundred pages!! (12mo), including the introduction. It will be difficult to get the volume within covers. A year's hard labor is bestowed upon it; I hope it will be useful and supply a desideratum. As a consolation for my honest faithfulness in making it tolerably thorough, and so much larger than I expected it would prove, it is now clear that I shall get nothing or next to it for my year's labor. At the price to which it must be kept to get it into our schools, etc., there is so little to be made by it, that I cannot induce a publisher to pay the heavy bills, except upon terms which swallow up all the proceeds; or at the very least I may get $200, if it all sells, a year or two hence.

Meanwhile, I have paid the expenses principally incurred on the first volume of "Illustrated Genera," which I can't print and finish till the "Manual" is out; have run heavily into debt in respect to these works, which were merely a labor of love for the good of the science and an honorable ambition; and how I am going to get through I cannot well see....

I should despond greatly if I were not of a cheerful temperament....

I wish I could write to you as you wish, all about botany, etc. I wish I could aid you as I desire, but I fear it is impossible. I must have rest and less anxiety. Two more years like the last would probably destroy me. If I had an assistant or two, to take details off my hands, I might stand it; as it is I cannot. Carey spent three months with me last season, and was to study and ticket your Texan collection in my hands, take a set for his trouble, and Mr. Lowell and Mr. S. T. Carey would take what they needed and pay for them, so that I could pay your book-bill at Fowle's. The utmost Carey found time to do was to throw the collection into orders; there they still lie, in the corner! There perhaps they had best lie, now, till the collection of the past season reaches me, when I will try to study them all together, along with Lindheimer's collections, a set of which still waits for me to study them. Will you wonder that I am a little disheartened when, in spite of every effort, I make so little progress? And in six weeks I begin to lecture in college again; and in April the Garden will require more time than I can give it. Such are merely some of the things on my hands, some of my cares! Still I am interested in you, and in your collections, and will do what I can....

Then if you will continue to send seeds (pretty largely), also bulbs, cacti, tubers, etc., now in early spring (and root-cuttings of some vines), taking pains that they are sent in a direct way, so as to come alive in May, etc., I will get an appropriation allowed from the Garden for you. Don't try other live plants till we have better communication with Texas. We have sunk money in this already and had to give it up....

Forgive my long neglect; accept my apologies. I'll see if I can do any better hereafter, when I have a wife to write letters for me.

March 10th.

Besides all the rest, the Academy's correspondence presses hard on me. I have written twenty-four letters for the steamer to-morrow. Fairly to keep up my correspondence and answer all my letters would take full two hours every day of the week except the Sabbath. So have mercy, and long patience....

Meanwhile my "Manual" is out; but not published till the 10th February. What can you expect from a man who takes up a job in February, 1847, to finish in May or June certain; but who, though he works like a dog, and throws by everything else, does not get it done till February comes round again. So it is only now that I have anything to send you. I am now printing off my "Genera Illustrata"—the text for one hundred plates; mean to have it out in a month; but I will not wait any longer....

TO GEORGE ENGELMANN.

CAMBRIDGE, February 29, 1848.

... Now for Fendler himself. He ought to go back, and without delay. He has gained much experience, and will now work to greater advantage. He makes unrivaled specimens, and with your farther instructions will collect so as to make more equable sets. If he will stay and bide his time he can get on to the mountains, and must try the higher ones, especially those near Taos.

Let him stay two years, and if he is energetic he will reap a fine harvest for botany, and accumulate a pretty little sum for himself, and have learned a profession, for such that of a collector now is. Drummond made money quite largely.

I had rather Fendler would go north and west than south of Santa Fé. New Spain and Rocky Mountain botany is far more interesting to us than Mexican.

TO JOHN TORREY.

March 29, 1848.

Your parcel came to-day; many thanks. After dinner I have just looked over the Mexican Compositæ of Gregg,[153] which are numerous, and quite a bonne bouche. My old love of the dear pappose creatures revived at the sight, and I longed to take them by the beard. If at liberty to do so (am I?) I think I will, at the same time I do the Santa-Féans; and at the same time I will study any of Abert's or Emory's Mexican or North Spain Compositæ you have not already disposed of. As to the parcel to be divided, of which there are no duplicates, whoever packed your parcel has taken care that there shall be pieces enough, if no specimens! They were in longer paper than the other bundles; not protected by binder's board, and therefore both ends, for two or three inches, were nicely bent up against the ends of the shorter bundle next them; which was very pretty for the shape of the parcel, but death to many of the plants; for the fold came just below the heads in most cases, too many of which were decapitated like the victims of the (last but two) French revolution.

I have been going on with recitations for some time, twice a week (two hours), and to-day I began my lectures to the whole Junior class, on Geographical Botany for the present.

What with these duties, superintending gardener, and painting and papering in the house, and Sprague drawing for the second volume of "Genera," and I printing the first, with the printer ever on my heels for copy, and at the same time printing Memoirs and Proceedings of the Academy, and managing large correspondence, you may conceive that my hands are full.

Yours most cordially,
A. GRAY.

TO W. J. HOOKER.

CAMBRIDGE, 2d May, 1848.

I send ... a copy (roughly put into paper covers) of the first volume of "Genera Illustrata," regretting there is not time to send you a bound copy. I hope you will like it. Sprague is improving fast, reads Brown's papers, etc., and is getting a good insight into structural botany, even the nicest points. We mean to carry on the work, and I hope for considerable London sale of it. The price is $6, or in London, £1 10s., which I trust will be thought low. Please notice it in the "Journal." The proceeds go principally to support Sprague in carrying on the work. I put his name on the title-page without his knowledge and at the expense of his great modesty.

I want to introduce the tussock grass on our eastern coast, where it will thrive well. Is it too late to send this spring? Or will you send in autumn?

P. S.—The last steamer brought good news of peace and strength in England, dissipating the alarm of many, but I felt none myself, having a strong confidence in the soundness of Old England and the durability of her institutions, of which I am here esteemed an over admirer.

Dr. Gray was married, May 4, 1848, to Jane L., daughter of Charles Greely Loring, a lawyer in Boston. In June they made a short journey to Washington, that Dr. Gray might, on undertaking to describe the plants of the United States Exploring Expedition, see Commodore Wilkes.

TO JOHN TORREY.

CAMBRIDGE, 8th May, 1848.

Yesterday I sent to Grant at Wiley's for you a parcel containing some "Linnæas," etc., received from Hamburg, your copy of Seubert on Elatine, and a bound copy of the "Genera Floræ Americæ Boreali-Orientalis Illustrata," which I ask you to accept, and which I trust you will like. There is also a specimen inclosed of some vegetable product that has lately become somewhat common here, and which I thought you might like to examine. It is apparently of a rather complicated structure, in fruit evidently, but syncarpous; the heterogenous and baccate or fleshy ovaries being immersed without apparent order in a farinaceous receptacle. If you should be at all puzzled. with it, and can't find out to what particular family it belongs, you might call in the aid of Mrs. Torrey and the girls, to aid in the investigation. I dare say you will make it out.

June, 1848.

I am just home this morning, and as I had no time yesterday to reply to your kind letter of Saturday, I write at once now....

Friday evening we were at the White house, to see Madame Polk. We have accomplished a great deal of sight-seeing and all in our week and a day, and J.

THE BOTANIC GARDEN HOUSE IN 1852

has enjoyed it much, except the drawback of not seeing Mrs. T. and the girls and yourself at home, which she greatly wished....

Now as to Exploring Expedition. We will talk it over in full when you come on here toward the end of this month.

Suffice it to say (as I am pressed for time) that I had made up my mind what I would do it for before I left home; that on looking over the collection, as to various parts of it, as far as time allowed, I found it less ample than I supposed, but with many difficulties owing to specimens in fruit only, or flower only. I think it no very awful job, if done in the way I propose, which is, not by monographs by people abroad, which the committee will not agree to, but by working up a part abroad in Hooker's, or Bentham's, or Garden of Plants herbarium.

The chairman of the committee and Wilkes behaved very well, and told me they were very desirous I should take it up.

On Friday evening Wilkes came in, before we went to the President's; asked me to say what I would do. I told him at once what I would do (just what I had told J. before we left Cambridge), and Wilkes at once accepted my terms, as I supposed he would. My terms were based on the supposition that there is five years' work in preparing for the press the collections left on hand, and in superintending the printing....

We must settle together the typographical form of the work, etc., when you come, and we will make the other writers conform to the plan we agree on, which perhaps you have already fixed.

Now I want a careful and active curator. What young botanist can I get?...

27th Nov., 1848.

Wright is up from Texas (with his mother at Wethersfield, Connecticut); he will soon be here as curator to me, taking Lesquereux's[154] place, who has been with me a little, but now, as a consequence of his visit to Columbus, goes to aid Sullivant, with a provision that makes the truly worthy fellow perfectly happy. They will do up bryology at a great rate. Lesquereux says that the collection and library of Sullivant in muscology are "magnifique, superbe, the best he ever saw."

TO GEORGE ENGELMANN.

January 24, 1849.

Halstead, I believe, has nearly decided to go on the Panama Railroad Survey; I trust to get Wright attached to the boundary survey. I have a letter from Fendler, in which he expressed his willingness to go to the Great Salt Lake country, if he can get government protection and food, etc. In a few days I shall write to Marcy; send him the sheets of "Plantæ Fendlerianæ," and make a vigorous application for this aid. No doubt I shall get it, I think. But perhaps it might be almost as well for Fendler to go over with a party of emigrants directly to Mormon City. But probably there will be emigrants bound for the same place, accompanying the regiment, as near as they go. Fendler can do admirably well in that region, if he perseveres. But will he not take the gold-fever and leave us in the lurch? Will not living, etc., be very dear in Mormon City also? I fear it. I must leave. much to your discretion. Only if you think Fendler has a strong tendency to gold-hunting (which few could resist) let him go. And afterwards, if he chooses to collect plants, very well. Few can withstand the temptation when fairly within the infected region, and we hear the Mormons have found gold also....

February 25, 1849.

I have just received from the secretary of war, Mr. Marcy, and inclose to you, what I think will procure all the facilities that Fendler can wish from United States troops. If, as I was informed, the secretary has no right to issue an order for rations to Fendler, he has certainly done the best thing by issuing a recommendation which will, if the commander is favorably disposed, enable him to give all without any order. Indeed, I think we could ask nothing better....

In my haste, and multitude of business, I have shabbily neglected to send the copies of "Plantæ Fendlerianæ" to Hamburg for Braun. And now the Danes have blockaded the Elbe....

I think I shall soon send the smaller things to you by express, and retain the three volumes of "Memoirs" for some opportunity less expensive. We want railroad all the way to St. Louis.

I am crowded—overwhelmed—with work. But college work will be over in July, and the second volume of "Genera," which I am now hard at work on, will soon be printed off; a week more and I shall have finished the copy.

I must then work at Exploring Expedition Compositæ, and soon at Fendlerianæ, and (when the sets arrive) at Lindheimer's, if you wish. I have made a genus of the Texan Rue—between Ruta and Aplophyllum,—e. g., Rutosma. I think there are some good remarks you will like in the second volume of "Genera."

I foresee an unusually good chance to get rid of the college work a year hence, and must therefore try to overhaul the Exploring Expedition plants, so as to get them into some shape, and next year (May or April) go abroad with them, sit down in London and Paris, and work them off. I will then drum up subscribers for Fendler and Lindheimer.

I want you to help me a little about Trees; our native trees up to Cornus inclusive, for this year, for the report I have promised the Smithsonian Institution.[155] I wish I had a good assistant; one who could work at botany. Perhaps I can find one abroad.

TO JOHN TORREY.

February 26, 1849.

Having determined on an expedition for Wright, you may be sure I was not going to be altogether disappointed. Accordingly I have got one all arranged (Lowell[156] and Greene subscribing handsomely) which is as much better than Emory's as possible, and thus far everything has wonderfully conspired to favor it. Wright has left me this morning to go to his mother's in Connecticut (Wethersfield); there to make his portfolios and presses; comes on to New York soon; takes first vessel for Galveston (I expect a letter from Hastings telling when it sails), and to reach Austin and Fredericksburg in time to accompany the troops that are about to be sent up, by a new road, across a new country, to El Paso, in New Mexico. Look on the map (Wislizenus) and you will see the region we mean him to explore this summer; the hot valley of Rio del Norte, early in the season, the mountains east, and especially those west in summer. He will probably stay two years, and get to Taos and Spanish Peaks this year or next. We shall have government recommendations to protection, and letters to an officer (commanding) who, through Henry, has already made overtures to collect himself or aid in the matter.

26th May, 1849.

I have finished all the copy of "Genera Illustrata," vol. ii., at length; the printer has yet two or three sheets to set up. The plates are working off in New York. It will now soon be off my hands. It is long since I have done anything at Exploring Expedition plants. I am now going at them. It is a shabby, unsatisfactory collection....

CAMBRIDGE, November 2, 1849.

... Sorry I am that you could not be here while Harvey is here; he will he south by Christmas. He desires me to say that he expects to spend the first half of December in New York at Dr. Hosack's, and will be most glad to see you. I am sure you will like him. We are perfectly charmed with him. A quiet, unaffected, pleasant man—extremely lovable. He works away at a table in my study. His course is a very interesting one. He is a beautiful writer, but not very fluent extempore, though with more practice he would be a fine lecturer. He has a good audience....

Sprague has promised now to take up and finish your quarto drawings. He says he can work but a little while at a time, from a difficulty of breathing. Had I foreseen his health and vigor giving way, I should not have undertaken the Trees, which, as to illustrations (as he is more fond of them than of anything else, and has made fine drawings), we have gradually enlarged our ideas about them much beyond the original plan, as to the figures. He must get this volume off his hands this winter, anyhow. The "Genera" will lie in abeyance....

My plan is only to bring out one volume of the Tree-Report next spring, and not to go beyond the limits of the United States proper, those of "Genera Illustrata," except to mention the trees of the far West in a general way; otherwise it would be far too formidable....

Sir John Richardson dropped in on me the very day Harvey arrived, expressed regret he could not see you, learned here the rumored news of Franklin. I wish you could have been here at a little dinner party we made for them. He is at home by this time.

TO W. J. HOOKER.

December 3, 1849.

... We are glad to hear what fine discoveries your son is making in Thibet, etc.

I saw to-day for the first time, at Green's, the Himalaya Rhododendrons....

I have just parted from Harvey, who has passed seven weeks with us, and having finished his course at the Lowell with much acceptance now joins his friends at New York and Philadelphia till Christmas, and then goes south to Florida, Alabama, and probably either to Jamaica (where Dr. Alexander now is) or to the mountains at the St. Iago end of Cuba, a terra incognita nearly. Harvey is a most winning man; my wife and I have become extremely attached to him, and are sorry to part with him.

We do not mean to let any naturalist be idle who comes to this country, so he is already engaged to give illustrations of our peculiar Algæ for the Smithsonian contributions and to prepare (after his return home, of course) a manual of United States Algæ after the fashion of his second edition of "British Algæ." There will be no small demand for it....

P. S.—Mr. Wright got through to El Paso in southern New Mexico, and is on his way back, with, he says, a fine collection.

We got some fine daguerreotypes of Harvey, so much better, he says, than he has seen in England that he has had an extra one taken for Lady Hooker.

TO GEORGE BENTHAM.

CAMBRIDGE, January 7, 1850.

Your letter of December 4th and your very flattering article in the December number of "Hooker's Journal" were both most gratifying; and the remarks on the Mimosa were timely, as I was just about consigning the manuscript of the earlier part of the new "Plantæ Lindheimerianæ" to the printer. I like what you say about "deduplication" much, and freely accept almost all. I took the name coined to my hand, not feeling at liberty to coin a new one. I think the production of new organs one before the other can be pretty well explained morphologically and anatomically, in accordance with your hint, and shall attempt to work it out in the third edition of my "Botanical Text-Book," which I am now preparing for the press. I shall be most glad of any further hints.

May I ask you what you think of Adrien de Jussieu's way of explaining the regular alternation of organs in the flower? I greatly incline to it....

I have to finish this Lindheimer collection, finish Fendler's, distribute and study Wright's collection when I get it, carry the "Botanical Text-Book" through the press, rewriting and expanding it (thus far I have made it all over), write the first volume of an elaborate report on the Trees of United States for the Smithsonian Institution, in fact a Sylva, with colored plates by Sprague (which I could not resist taking in hand, as that institution promised to bring it out, and handsomely, at their expense), and give my course of lectures in the college from March to June. When all this is done I can cross the Atlantic.... By engaging a brother professor to take the duties which I have for the autumn term (assigning to him pro rata from my salary), I shall be free until 1st March ensuing. But I mean to ask for leave of absence for a year, and trust I shall get it....

As far as it has yet shaped itself my plan is ... to sit down hard to work for the autumn and winter on the Exploring Expedition plants, to go to Paris in the spring and settle such questions as must be settled there after I come to know better than I now do (except in the Compositæ) what they are. Excepting the Oregon and Californian plants, which are assigned to Torrey, and the Sandwich Islands Collection, a fine one, the collection is a poor one, often very meagre in specimens, too much of an alongshore and roadside collection to be of great interest. I am not familiar with tropical forms and have no great love for them. I dislike to take the time to study out laboriously and guessingly, with incomplete specimens, and no great herbaria and libraries to refer to, these things which are mostly well known to botanists, though not to me, and I want to be taken off from North American botany for as short a time as possible. I must therefore come abroad with them, which the pay that is offered will enable me to do. I have found a good deal to interest me in the Compositæ, especially those of Rio Negro, of north Patagonia and of the Andes of Peru....

Now, will you take it as a bore, an imposition on your kindness, if I frankly ask whether I can possibly offer you any sort of inducement to aid me, at least so far as to run over the collections with me, and name those that are familiar to you as we pass, and refer others, as nearly as one can without study, to their proper places? Your mere comments in running through would save half my time.

It is most natural that you should not incline to any such trouble, and I know your hands are always full; so, if you say no, I shall feel it is quite right, and do the best I can....

We shall be most glad to visit you at Pontrilas house at whatever time best suits Mrs. Bentham and yourself, whether in summer or in autumn, any time before we settle down into our winter quarters....

With best wishes to Mrs. B. and yourself for the new year, I am very faithfully yours,

A. GRAY.

TO W. J. HOOKER.

April 2, 1850.

We were most glad to receive your kind favor of January 29, which, however, lay over a fortnight in England, and in the mean time we heard not only of Dr. Hooker's capture, but also, with much gratification, of his release. What an indefatigable man he is!

Finding myself greatly behindhand, on account of various hindrances and miscalculation of time, and utterly unable to accomplish half the work I had intended to do this spring, I have decided to break off; and to sail, in a packet-ship from Boston, on the 5th of June, with Mrs. Gray, for Liverpool, which we may hope to reach by the close of that month. This will give us an opportunity of seeing England in its summer dress, and to make, almost immediately following the sea voyage, a trip up the Rhine to Switzerland. On our return I must set to work diligently, and for a little while with Mr. Bentham, who has kindly offered to look over the tropical collections, which I know little of, and love as little.

The rewriting of all the structural parts of my 3d edition of the "Botanical Text-Book," which I was inadvertently drawn into, has proved a most time-consuming business. It is not yet through the press.

Wright's collection of seeds I had divided into two parts, and I send you one by the hands of Mr. Lowell, who with his whole family goes out by this steamer. You will receive them in good time to raise them....

Mr. Lowell is of great use to us, in helping on these explorations, and I look to his visit to Europe, the sight of the great collections, and the society of naturalists to strengthen his tastes and fire his zeal in these respects.

I long to have him and Mrs. Lowell, a very good. friend of ours, make your acquaintance.

FOOTNOTES:

[1] This colony was composed of rigid Presbyterians, who desired to leave Ireland to escape various persecutions. They sent out the Rev. Mr. Boyd, early in 1718, with an address to the Governor of Massachusetts. The address, now in the Archives of the New Hampshire Historical Society, was signed by three hundred and nineteen persons, nine of whom were clergymen. The report brought back by Mr. Boyd of his reception by the governor and of the prospects of the country was so favorable that the addressors converted their property into money, and embarked in five ships for Boston, which they reached August 4, 1718. In Boston they separated for different places, but the larger part were sent to Worcester, then a frontier settlement of fifty-eight dwellings and two hundred inhabitants, but needing a larger population as protection from the Indians. John Gray—there were two of his name in the original party—went to Worcester, where he owned considerable land, and was evidently a man of influence in the colony, to judge from the various public offices held by him.

[2] Robert Gray, one of John Gray's sons, was twenty years old when he came to America. There is a tradition in the family that the acquaintance and courtship began on the voyage.

[3] Sauquoit was a settlement in the eastern part of the town of Paris, the township so named in grateful recognition of a supply of food, sent by a Mr. Paris, of Oswego, at a time when the early settlers were near starving.—A. G.

[4] Moses Gray was the eighth child,—a boy and a girl were born later,—and one step-brother, Watson, survived Moses Gray. Moses Wiley Gray made the journey to Sauquoit, on horseback, taking before him his son Moses, then a boy of eight. The Mohawk Valley at this time was the far West, with only slow and tedious communication beyond Schenectady, but opening, in its lovely tributary valleys, tempting regions of hill and valley, well wooded, with clear, sparkling streams. The land offered good farming opportunities when cleared of trees, and the rapid streams gave good promise of water power. Here Moses Wiley Gray took a farm on the top of a hill, and cultivated the land for ten years. He was injured by the fall of a tree, and his leg was amputated. He died the next day, May 8, 1803, leaving his son Moses, with his stepmother and her children largely dependent on his assistance.

[5] She was married in 1788.

[6] The house is still standing which, built in 1648 by an ancestor of Ralph Waldo Emerson, was bought by William Howard, in 1669.

[7] Asa's mother was but four years old, when the family moved to Sauquoit, and well remembered her mother's crying at the crossing of the Hudson River, which must have seemed formidable in the small boats of that time. Joseph Howard was a man of a very lovable character, as shown from the affectionate remembrances of him by his grandchildren, the eldest of whom, Asa, was much with him. He was a deacon of the First Church in Sauquoit for forty years, and one of the leading men in the town. He died in 1849.

[8] Moses Gray was a man of great activity and energy. He soon added a shoe-shop to his tannery, where he hired a few hands to make shoes from the hides he tanned, taking these again by wagon to Albany, a journey of many days, where he bought his skins and some necessary supplies. Money was scarce in the newly settled country, and the things needed were mostly got by exchange. Meantime, as the chance came, he was buying land on the hills around. Clayville is where the valley narrows towards the source of the Sauquoit Creek, as "rivers" are called in that neighborhood in old Dutch fashion, and the hills are sharper and rougher. The scenery, however, is still beautiful, and the house which Moses Gray built two or three years later yet stands, with a lovely near view of stream and hill and wood. Asa Gray remembered his father building it. Busy as the father was out of doors, the mother was perhaps busier still. Asa, the younger brother by the first wife, was dying of consumption; he was moved on a bed from Sauquoit to Paris Furnace, and died very soon after, in May, 1811, aged twenty-three. When the child was born, November 18, 1810, it was carried to him to see, and he said he wished they would call it Asa, if it had had no name as yet decided on. He was of a singularly sweet and gentle character. The step-brothers were taken in turn to be taught and trained. The hands employed on farm or in trade were generally lodged and boarded. Often their clothes were mended or made. The wheat and grain were home-raised, as were all the vegetables. There was little fresh meat, except when a sheep or beef was killed, and that meant salting and curing. Butter and cheese were all homemade, and could be taken to Albany for sale, as was also grain; as the farm grew, more cows were added. Then the clothing was homemade. The wool for flannel sheets and underclothing and for the men's clothes was home-spun, the nicer portions taken off and carded separately, and spun as worsted for the children's and women's dresses; also the yarn for socks for the whole family. A spinning-girl was hired for part of the year, for flax was also spun for the house linen and for wearing-apparel. The weaving was hired out. The tailor came by the week to make up the clothing with the mother's help, and after the tannery was

given up, the shoe-maker came at intervals to make the shoes. As the girls grew older they took their share at the wool and flax wheels. It is said that the first spinning of flax on the small wheel was introduced by the party of Scotch-Irish emigrants of 1718; that the women gave lessons to the women of Boston on Boston Common, and the fashion was so set for that spinning. It is also said that the Irish potato was first introduced into New England by these same colonists.

A widowed sister came with her children to make her home under the same roof when the Grays moved later to a larger farm, and there seemed always some boy to be housed and taught and trained. Though his aid might tell out of doors, the home care came upon the mother. But Mrs. Gray was a woman of singularly quiet and gentle character, with great strength and decision, and possessed a wonderful power of accomplishing and turning off work; a woman of thoughtful, earnest ways, conscientious and self-forgetting.

The father was quick, decided, and an immense worker; from him the son took his lively movements and his quick eagerness of character, perhaps also his ready appreciation of fun.

[9] His mother, having another child, was probably glad to have the active boy safe for a few hours. Her young sisters lived not far away, and the youngest aunt, a girl of ten, was proud to take him to school; she had already taught him his letters. His father promised him a spelling-book of his own as soon as he reached *baker*, which was a marked spot of advance in the spelling-book. A few weeks saw him far enough on, and the coveted prize was given. He went proudly to school the next day, and as he could not speak to the teacher to proclaim his triumph, he walked in front of her desk to his seat, waving the book with a great flourish before her! It was just before he was three years old.

[10] Of one of these, his friend, now over eighty years old, gives an account in the succeeding letter:—

SAUQUOIT, *February 19, 1888.*

DEAR A.—I would like to give you some information of your uncle's early life if I were well informed, but, I have only one little incident, and perhaps that would be of small account at the present era, though at the time it took place it was of great moment to us both as children. Asa lived with his parents at Paris Furnace, now Clayville. I lived where Mr. Bragg afterward built his new house. Well, we had a lovely teacher that summer by the name of Sally Stickney, living at Colonel Avery's. She ruled by gentleness. For our class she had an old-fashioned two-shilling piece, with a hole through to insert a yard of blue ribbon. She put this over the head of the one that stood first in our class. So it traveled every night, all that summer, with some one of us, until the ribbon was worn and faded. But more than all that, the one that stood at the head on the last day of school was to be the owner of that two-shilling piece that we had watched with jealous eyes so many weeks, and studied Webster's old spelling-book so hard to gain. I think our eyes must have magnified it, for I have never seen a coin since that seemed so large. I think it was the same in Asa's eyes. Well, with hearts beating fast, and eyes on the coveted prize, we were called on the last day of school to spell; we took our places; I was at the head, Asa next. I missed and he went above me; my all was gone, but it was worse to have him point his finger at me and say out loud "kee-e-e." I braved it without a tear; a few more words would end the strife. It came around to him, and he missed; how quick I went above him; but in an instant he dropped his head on the desk before him and wept as though his heart would break. School was dismissed, scholars were leaving; still he did not move, until our kind teacher came to him, whispered to him, soothed and petted him; then he jumped up and ran, I suppose wishing me in Halifax. I felt sorry for him and would have been willing to divide with him if he had not crowed over me so. I ran nearly all the way home—a good mile—with my treasure, in great haste to have some one tell me the best way to invest my money. I was told to go another three quarters of a mile to Stephen Savage's store, spend it for calico, piece it up, to keep forever. I could get only one yard for my two-shilling piece, not nearly as good as can be bought now for three cents a yard. Not a trace of the quilt is left, nor of the old schoolhouse, or of those merry children; perhaps a few have wandered on to fourscore years. So it is little I can relate of his childhood, as the next year we moved from that district, but as years passed on I often heard of his rising fame with pleasure. If Eli Avery were living he would have been his best biographer in this place.

The time has flown so fast since all this transpired, it seems as if his tears had hardly dried before my grandchildren were studying his Botanies.

Two years ago the 9th day of September, when the doctor was visiting in Sauquoit, he called here and remarked, in his smiling way, "that he had got all over feeling badly about *that*." I said, "And well you may when you have received so many honors since then."

Your loving friend,
HARRIET ROGERS.

A neighbor who survived to a great age also told a story of Dr. Gray's boyhood, which he said he had from Dr. Gray's father:—

One day he had been set to hoe a certain amount of corn, and his father found him reading instead of at his work. He gave him his choice, to finish his task and then read comfortably, or to sit there in the field all day in the hot sun, which one knows is no pleasant thing in August, and read. He chose the reading, and his father said then, "I made up my mind he might make something of a scholar, but he would never make a farmer!" And so his farther education was decided.

[11] Asa Gray was the eldest of eight children, three sisters and four brothers, of whom there survive two sisters and two brothers.

[12] Dr. Gray visited Fairfield again in the summer of 1860 or 1861. He pointed out his old room, and told about some of the pranks he and his room-mate Eli Avery had played there as boys, especially once when they barred their room, escaped through the window by clambering down a rope, and then enjoyed the efforts of the master to break the door down. Oddly enough there was then a fresh panel in the door, as if a later generation had tried the same trick. There were a great many stories told of his exploits as a boy. But he said everything had been fathered upon him, and that few were really true. He was no doubt restless and active, and learning quickly and easily would have leisure for some mischief, but he said, "I always learned my lessons." He loved to recall the long rambles through the woods on Saturday holidays, and how in early spring he and his companions would climb to a lookout and see where columns of smoke could be seen above the trees, and so aim for the spot where they were making maple sugar. There they would beg a little syrup, and, boiling it down over their own fire and cooling it on snow, make a candy more delicious than any confectionery of after life. He remembered how he trained himself to know the trees by their bark as he ran through the woods, without looking up at the leaves, having then the keen power of observation though no especial interest in botany. For, as he always said, his first fancy was for mineralogy rather than for botany.

And he told how when he was a medical student, as so many about him were smoking, he tried it too; it made him very sick at first, and took him some time to get accustomed to it. At last, as he sat one evening before the fire and smoked, he said to himself, "Really, I am beginning to like it. It will become a habit; I shall be dependent upon it." And so he threw his cigar into the fire and gave up smoking entirely.

[13] In later life the novels were always saved for long journeys. The novel of the day was picked out, and one pleasure of a long day's ride in the train was to sit by his side and enjoy his pleasure at the good things. The glee and delight with which he read Hawthorne, especially the *Wonder-Book* and *Tanglewood Tales*, make days to remember. So he read George Eliot, and Adam Bede carried him happily through a fit of the toothache. Scott always remained the prime favorite, and his last day of reading, when the final

illness was stealing so unexpectedly and insidiously on, was spent over *The Monastery*, which he had been planning to read on his homeward voyage in 1887.

[14] It was established as a college in 1812, having existed as a school in the academy since 1809. There were then only five others in the United Stated: Philadelphia, New York, Boston, Dartmouth, and Baltimore. The war of 1812 with Great Britain made a demand for army surgeons along the frontier, and New York and Boston were too far to send the young men to be educated. Dr. Hadley was professor in the literary academy, and Dr. Willoughby, who had a wide medical reputation, was also in Fairfield. They planned a medical college, and applied to the legislature for aid; the sum of $5,000 was granted, and later, in 1812, $10,000. The first Faculty was organized by the Board of Regents of the New York State University, which had control of the educational institutions in the State. It grew rapidly in favor, and soon outnumbered the schools of the large cities. In 1820 the school had one hundred students, and increased to two hundred and seventeen later, and was the largest medical school in the country, except the one at Philadelphia. After the Albany and Geneva medical schools were established, it was seen there was no need of three so near together, and Fairfield Medical College was discontinued in 1840. In the list of graduates of Fairfield Academy were Albert Barnes, the noted expositor, General Halleck, of the United States Army, and James Hadley, professor of Greek at Yale and the distinguished linguist. In the records of the academy it is stated that "Asa Gray entered Fairfield Academy in the fall of 1825, and at the second weekly meeting joined the Calliopean Society of the institute. His handwriting on the register is still preserved, as well as all his doings as a boy while here, since he entered at an early age, being in fact much younger than the majority of the students." He graduated from the medical college January 25, 1831 in a class of forty-four. His rank was seventeen in the class on graduation. The subject of his thesis was "Gastritis." Two old catalogues are preserved at Fairfield. In the first there is the programme of studies at the academy for the year 1826; the other, dating January, 1832, contains a list of the professors of the medical college, the cost of instruction, and the outlines of two courses of lectures. One of them was given by Dr. Mather, who was a fellow student of Asa Gray's, and who still, at over eighty, retains a lively recollection of the eager, active young man whom his friends already thought would make his mark in the world; the other by Dr. Gray himself. This was one of the first courses of lectures which he delivered. The ticket-fee was four dollars. He kept through life a certain love for medicine and surgery, and a lively interest in its science and progress. These old studies and the country practice he had with the physician who was always his good friend, Dr. Trowbridge, often

served him on his journeys, when a regular practitioner was not within easy reach.

[15] Amos Eaton, 1776-1842. Graduated from Williams in 1799. Teacher, lecturer, and author of *Manual of the Botany of North America*, as well as of many reports on geological surveys.

[16] College catalogue of Fairfield, 1830-31.

[17] Lewis C. Beck, 1798-1853; professor in Albany Academy; author of *Botany of the United States North of Virginia.*

[18] Lewis David Schweinitz, 1780-1834; the first American who studied and described the fungi of the United States. He wrote also on other North American cryptogams and carices.

[19] John E. Le Conte, 1784-1860; formerly major in United States army. His first botanical publication was a catalogue of the plants on the island of New York, in 1810. He later wrote chiefly on entomology.

[20] It appears that in December, 1834, I read to the Lyceum of Natural History my first paper, *Monograph of North American Rhynchosporæ*, and my second, *New or Rare Plants of the State of New York*. They must have been printed early in 1835.—A. G.

[21] J. G. C. Lehmann, 1793-1860; professor at Hamburg.

[22] George Francis Reuter, 1815-1873; directer of the Botanical Garden at Geneva; curator of Boissier's herbarium.

[23] Edward T. Channing; professor of rhetoric and oratory at Harvard University.

[24] Daniel Treadwell; professor in Harvard University of applied physics; distinguished inventor in mechanics, especially in the welding of steel.

[25] John H. Redfield; curator of the herbarium of the botanical department of the Philadelphia Academy of Natural Science.

[26] *American Jour. Sci.*, xxv. 346-350.

[27] Dr. Trowbridge.

[28] *North American Gramineæ and Cyperaceæ*, of which Part I. was issued in 1834, Part II. in 1835. This was the first separate and individual publication by Dr. Gray. Sir W.J. Hooker said of it:—[It] "may fairly be classed among the most beautiful and useful works of the kind that we are acquainted with. The specimens are remarkably well selected, skillfully

prepared, critically studied, and carefully compared with those in the extensive and very authentic herbarium of Dr. Torrey."

[29] Alluding to the then popular squib of Major Jack Downing's letters.

[30] S. Wells Williams, 1812-1884. Went as missionary to China in 1833. Wrote a Chinese dictionary and other works; translated Genesis and Matthew into Japanese also. Later was secretary of the American Legation to China; returned to America in 1875.

[31] *Elements of Botany.*

[32] Chester Dewey, 1784-1887; professor in Williams College, Massachusetts. Removed to Rochester, N.Y., 1836, where he died. "Carried on the study of Carex and published on them for more than forty years" [A. G.].

[33] Jacob Whitman Bailey. 1811-1857; professor in the Military Academy at West Point. One of the earliest students of American Algæ, and distinguished also for his microscopic researches in botany.

[34] Charles W. Whipple, died in 1855. Was educated at West Point, where probably he was a pupil of Dr. Torrey. He was never in the army, but studied law and practiced in Detroit; was made Judge, then Chief Justice of the Supreme Court of Michigan. Ex-officio regent of the State university.

[35] David Bates Douglass, 1790-1849. He held the professorship of natural philosophy and civil architecture in the University of New York, and was afterward president of Kenyon College. He laid out Greenwood Cemetery.

[36] Dr. Thomas Raffles; a distinguished Congregational clergyman in Liverpool from 1812 to 1863.

[37] John Shepherd, b. 1764. For thirty-five years at the Liverpool Botanic Garden.

[38] Peter D. Knieskern, M. D., 1798-1871. "Botanized over the pine-barrens of New Jersey with utmost assiduity and skill, a simple-hearted, unpretendingly good and faithful man.... Few botanists have excelled him in their knowledge of the plants of the region in which he resided, and none in zeal, simplicity, and love of science for its own sake."—A. G.

[39] Thomas Nuttall, 1784-1859; a great traveler and explorer. Came to the United States in 1807. His writings are intimately connected with the development of North American botany.

[40] Robert K. Greville M. D., 1794-1866; author of *Scottish Cryptogamic Flora*, *Flora Edinensis*, and *Algæ Britannicæ*.

[41] Robert Graham, M. D., 1786-1845; professor of botany in the University of Edinburgh.

[42] William Nicoll. Invented section-cutting of recent and fossil woods in 1827.

[43] James Forbes, 1809-1861; professor of natural philosophy in the University of Edinburgh.

[44] Robert Jameson 1774-1854; professor of natural history in the University of Edinburgh.

[45] John Hutton Balfour, M. D. 1808-1885; professor of botany in Glasgow, and afterwards in the University of Edinburgh.

[46] Sir Charles Bell, 1774-1842; a very distinguished surgeon; author of *Anatomy of Expression* and many celebrated works. He accepted the chair of surgery at Edinburgh, 1836.

[47] John Abercrombie, M. D., 1781-1844; celebrated Scotch physician and author.

[48] Sir Robert Christison, 1798-1882; professor of materia medica in the University of Edinburgh.

[49] James T. W. Johnston, 1796-1865; agricultural chemist; professor at Durham. Lectured in the United States.

[50] Francis Boott, 1792-1863. Born in Boston, United States. Early removed to London, where he studied and practiced medicine a few years. "A good botanist, and in his later life devoted to the study of Carices" [A. G.].

[51] John Joseph Bennett, 1801-1876; keeper of the herbarium of the British Museum. "One of the most learned and modest of men" [A. G.].

[52] Sir John Richardson, M. D., 1787-1865. "The well-known Arctic explorer, zoölogist, and botanist" [A. G.].

[53] David Don, 1795-1856; librarian of the Linnæan Society; professor of botany in King's College, London.

[54] Frederic Pursh, 1774-1820. Emigrated to America, 1799. Traveled and collected much; settled later in Montreal, where he died.

[55] Aylmer Bourke Lambert, 1762-1842; author of the *Genus Pinus* and the *Genus Cinchona*. Owned a very large herbarium comprising plants of Pursh, who published under his liberal patronage.

[56] John Forbes Royle, M. D.; a surgeon in the East India Company. Wrote on the botany of the Himalaya.

[57] John Edward Gray, 1800-1875; keeper of the zoölogical collections of the British Museum for many years. "Of persistent ardor, indomitable energy, and great practical power" [A. G.].

[58] William Clift, 1775-1849; curator of the Hunterian Museum of the Royal College of Surgeons.

[59] Peter Mark Roget, M. D., 1779-1869; secretary of the Royal Society, London. Wrote *Animal and Vegetable Physiology*, and the well-known *Thesaurus.*

[60] Richard Taylor; printer; for many years secretary of the Linnæan Society.

[61] Francis Bauer; botanical artist to George III.

[62] Thomas Horsfield, M. D., 1774-1859. Born in Pennsylvania. After sixteen years in Java, passed the rest of his life in London as keeper of the museum of the East India Company. Brown & Bennett published part of his collections, *Plantæ Javanicæ Rariores.*

[63] I forgot to mention also some bricks from Babylon, covered with arrowhead characters, which were the most interesting relics of antiquity I almost ever saw.—A. G.

[64] Nathaniel R. Ward, 1791-1868; inventor of the Wardian case.

[65] Archibald Menzies, 1754-1842; the botanist who accompanied Vancouver in his voyage to the west coasts of North and South America. His collections are in the Edinburgh and Kew Herbariums.

[66] Sir Charles Lyell, the geologist.

[67] Edwin J. Quekett, 1808-1847. Wrote much on the microscopic structure of plants and animals.

[68] John Miers, 1789-1879; a botanist who studied in South America and wrote many papers.

[69] Thomas Walter, d. 1788, in Carolina, U. S. Wrote *Flora Caroliniana.*

[70] James Scott Bowerbank, 1797-1877. Wrote on Sponges and the Fossil Fruits of the London Clay.

[71] Edward Forster, 1765-1849. Made vice-president of the Linnæan Society in 1828.

[72] Herbert Gray Torrey, born just before Dr. Gray sailed, was his godson.

[73] Mr. George P. Putnam; the American publisher and bookseller, at this time established in London.

[74] William Valentine, a very promising young botanist, who wrote valuable papers on the structure of mosses. Went early to Tasmania, where he died.

[75] Pulled down in 1891.

[76] Theodore Hartweg, died in 1871. Explored in Mexico and California, 1836 to 1847; later director of the Grand-ducal Gardens, Swetzingen, Baden.

[77] Philip Barker Webb, 1793-1854; a "distinguished English botanist residing in Paris, of vast and varied knowledge. He accumulated one of the largest herbaria, bequeathed to the Duke of Tuscany."—A. G.

[78] Beaupré Charles Gaudichaud, 1780-1854; French botanist. Went round the world in the Bonite, and published the Botany of the expedition.

[79] Jacques Gay, died 1863. Born in Switzerland, and a pupil of Gaudin.

[80] Jean F. Camille Montagne, 1794-1865; surgeon in the French army. Retired in 1830, and devoted himself to cryptogamic botany.

[81] Achille Richard, 1794-1852; professor of botany in the Ecole de Médecine, Paris; son of L. Claude Richard.

[82] Edouard Spach, 1801-1879; native of Strasburg, many years keeper of the herbarium at the Jardin des Plantes.

[83] Charles François Brisseau Mirbel, 1776-1854; one of the most distinguished vegetable anatomists of the age. His earliest publication in 1801.

[84] Auguste de St. Hilaire, 1779-1853. Accompanied the Duke of Luxembourg on his voyage to Brazil, where he spent six years, and published a *Flora of Brazil*, 1825, and many other works.

[85] Baron Benjamin Delessert, 1773-1847; a French financier and philanthropist. Associated with De Candolle in the publication of the *Icones Selectæ*.

[86] Jean Louis Berlandier, died 1851; a Belgian. Established as an apothecary at Matamoras, 1827 or 1828. The first botanist to explore New Spain. He also made large collections in western Texas.

[87] Adolphe Theodore Brongniart, 1801-1876; distinguished French botanist, more especially in fossil botany; professor of botany at the Jardin des Plantes.

[88] The rediscovery of Shortia in 1878 is described on p. 682.

[89] Nicolas Charles Seringe, 1775-1856; professor at Lyons. Seringia named for him.

[90] Esprit Requien,1788-1851; a pupil of A. P. de Candolle at Montpellier. Often quoted in the *Flore Française.*

[91] Michel Felix Dunal, 1789-1856; professor of botany at Montpellier. "One of the earliest friends of A. P. De Candolle. Author of several important monographs" [A. G.].

[92] Alire Raffeneau Delile, 1778-1850; director of the Garden of Agriculture established at Cairo. Later he succeeded De Candolle in the Botanic Garden, Montpellier. A celebrated botanist.

[93] There is a gigantic statue of Columbus, placed in a conspicuous place and looking down into the harbor. They make very much of him now, as well they may; they derided him when living, they set up his image long after he is dead. Of course we are very much obliged to him, for if he had not discovered America what would have become of us!—A. G.

[94] Gaetano Savi, 1769-1844.

[95] Antonio Targioni-Tozzetti, 1785-1856; distinguished Florentine botanist.

[96] Giovanni Battista Amici, 1784-1863; an Italian astronomer, especially skilled in the construction of optical instruments.

[97] Horatio Greenough; the American sculptor in Florence.

[98] Roberto de Visiani, 1809-1878; professor of botany at Padua; author of a *Flora Dalmatica.*

[99] B. Biasoletto, M. D., 1793-1858. Triest. "A botanist of merit and investigator of Algæ of the Adriatic" [A. G.].

[100] M. J. Tommasini, 1794-1879. Triest. Author of a Botany of Mt. Slavonik, Istria.

[101] Stephen Ladislaus Endlicher, 1804-1849; professor of botany in the University of Vienna; author of *Genera Plantarum.*

[102] Aloys Putterlich, 1810-1845; keeper of the Botanical Museum, Vienna.

[103] Edward Fenzl, 1807-1879; professor of botany and director of the Botanic Garden at Vienna.

[104] Dr. Heinrich Schott, 1794-1865; director of the Imperial Gardens, Schönbrunn. "He was the highest authority on Aroideæ" [A. G.].

[105] A. C. J. Corda, 1809-1849. Prague. A distinguished mycologist. Lost at sea on returning from America.

[106] Karel B. Presl, 1794-1852; professor at Prague and curator of the herbarium.

[107] Christian Gottfried Nees von Esenbeck, 1776-1858; professor of natural history at Bonn and Breslau.

[108] Joseph Gerhard Zuccarini, 1797-1848; professor of botany at Munich. Among other publications he assisted in describing the plants collected and described by Siebold in the *Flora Japonica.*

[109] Julius Hermann Schultes, 1804. Died in Munich, 1840.

[110] Jean Etienne Duby, 1797-1885; long one of the Genevese clergy and a botanist and colleague of Augustin Pyramus do Candolle.

[111] Dr. Gurdon Buck.

[112] Hugo von Mohl, 1805-1872. Born at Stuttgart. Professor of botany at Tübingen. "Chief of the vegetable anatomists of this generation" [A. G.].

[113] Eduard Friedrich Pöppig, 1798-1868; professor of zoölogy at Leipsic. Made collections of plants in Cuba, Chili, Peru, and on the upper Amazon.

[114] Christian Friedrich Schwägrichen, 1775-1853; professor of natural history at Leipsic.

[115] Heinrich Gottlieb Reichenbach, 1793-1879; professor of botany at Dresden. A voluminous author, especially of illustrated works on European plants.

[116] D. F. L. von Schlechtendal, 1784-1866. University of Halle. Editor of the *Linnæa* and *Botanische Zeitung.*

[117] Christian Schkuhr, 1741-1811. *History of Carices*, 1802.

[118] Dr. J. H. Klotzsch, 1805-1860; keeper of the Royal Herbarium at Berlin.

[119] Karl Sigismund Kunth, 1788-1850. Appointed professor of botany at Berlin, 1819. Author of *Enumeratio Plantarum* and other well-known descriptive works.

[120] Christian Gottfried Ehrenberg, 1794-1876. Berlin. Student of the microscope, and author of works on the lower forms of plants and animals.

[121] Heinrich Friedrich Link, 1767-1851. Professor at Breslau, then at Berlin. Wrote *Anatomy of Plants* and *Elements of Botanical Philosophy.*

[122] *A Flora of North America*; containing abridged descriptions of all the known indigenous and naturalized plants growing north of Mexico; arranged according to the natural system. By John Torrey and Asa Gray. New York. 8vo; vol. i., 1838-1840, pp. xvi, 711; vol. ii., 1841-1843, pp. 504.

[123] S. Constantine Rafinesque-Schmaltz, d. 1840. A Sicilian by birth. First arrived in the United States, 1802, for three years; returned in 1815, and explored the Alleghanies and Southern States. "An eccentric but certainly gifted personage, connected with the natural history of this country for the last thirty-five years" [A. G.].

[124] Benjamin D. Greene, 1798-1862. First studied law; then medicine in Scotland and Paris. Devoted himself to botany. "His very valuable herbarium and botanical library were bequeathed to the Boston Natural History Society. He was always a most liberal and wise patron of science" [A. G.].

[125] Jacob Bigelow, M. D., 1787-1870; an eminent Boston physician; author of the *Floral Bostoniensis*, 1814.

[126] George B. Emerson, 1797-1881; an eminent teacher in Boston, Mass.; author of *Trees and Shrubs of Massachusetts.*

[127] Director of Kew Gardens.

[128] Ferdinand Lindheimer, 1801-1879. Died at New Braunfels, Texas. A German. "An assiduous and excellent collector and a keen observer; his notes, full and discriminating, add not a little to the value of the collections" [A. G.].

[129] Edward Tuckerman, 1817-1886; professor at Amherst. "The most profound and trustworthy American lichenologist of the day" [A. G.].

[130] Benjamin Peirce, 1809-1880; professor of mathematics, Harvard University.

[131] Carl Geyer, 1809-1853; a German botanist who explored the basin of the upper Mississippi with Nicollet under the Bureau of

Topographical Engineers, 1836-1840. Afterwards crossed the Rocky Mountains to Oregon.

[132] Lecture to his class in college.

[133] Dr. Gray imported a quantity of small evergreens from England and planted the ground extensively, adding also many other kinds.

[134] Dr. Rugel came to America, 1842; settled in eastern Tennessee and collected in the southeastern States.

[135] To his college class.

[136] Dr. Jeffries Wyman.

[137] This was Dr. Gray's second course of Lowell lectures. Dr. John A. Albro, the Congregationalist minister of Cambridge, was his pastor.

[138] G. J. Mulder, 1802-1880; professor of chemistry in the University of Utrecht. Wrote on Animal and Vegetable Physiology.

[139] Carl H. Schultz-Schultzenstein, 1798-1871; professor of physiology in the University of Berlin. Wrote voluminously upon Cyclosis and the Vessels of the Latex, etc.

[140] Dr. Charles Danbeny, G. B., 1795-1867; professor of botany and rural economy at Oxford; chemist and geologist.

[141] William Oakes, 1799-1848. "The most thorough and complete collector and investigator of New England plants" [A. G.].

[142] Alphonse Wood, 1810-1881; author of popular botanical text-books.

[143] His brother, then living with him in Cambridge to enter Harvard.

[144] Christian Hendrik Persoon, 1755-1838; a botanist at the Cape of Good Hope. Died in Paris at a very advanced age. Fungologist.

[145] The third course of Lowell lectures.

[146] Augustus Fendler, 1813-1883. Came from Prussia to America in 1840. Collected in New Mexico, and on the Andes about Tovar in Venezuela, and in Trinidad. "A close, accurate observer, a capital collector and specimen-maker; his distributed specimens are classical. Of a scientific turn of mind in other lines than botany" [A. G.].

[147] Spencer F. Baird, afterward widely known as secretary of the Smithsonian Institution.

[148] A relative left Nuttall a comfortable little estate and property on condition that he should not be away from it more than three months in the year. He managed to come to America again by taking the three last months of one year and the three first of the next.

[149] Increase Allen Lapham, 1811-1875; author of a Catalogue of Plants in the Vicinity of Milwaukee.

[150] Charles Pickering, 1805-1878. "Author of *Geographical Distribution of Plants and Animals and Man's Record of his own Existence*, largely a record of changes in the habitat of plants. A monument of wonderful industry" [A. G.].

[151] John P. Brace, Litchfield, Conn.; an early botanist and mineralogist. His herbarium went to Williams College.

[152] A. Wislizenus, M. D., b. 1810. Explored New Mexico and Mexico; was arrested as a spy. On returning to the United States published a memoir of the tour, 1846-1847.

[153] Josiah Gregg, died in California, 1850; made excellent collections in Chihuahua and in the Valley of the Rio Grande. Author of the *Commerce of the Prairies.*

[154] Leo Lesquereux, 1806-1889; the leading fossil botanist of America, and a distinguished bryologist.

[155] Sprague made, under Dr. Gray's directions, some drawings in color of the work planned, *The Trees of North America.* The work was never completed, too many things, expense, etc., coming in the way, but the few plates printed and colored by Prestele were issued in a small quarto pamphlet by the Smithsonian Institution in 1891.

[156] John Amory Lowell, 1798-1881; a Boston merchant, and a liberal patron of botany. He bought many valuable books and collected a fine herbarium. He shaped the policy and direction of the Lowell Institute founded by his cousin, John Lowell.

Lightning Source UK Ltd.
Milton Keynes UK
UKHW010638090123
415051UK00005B/420